Maths IN THE Real World

On the BUILDING Site

Faye Cowin

NELSON
CENGAGE Learning

Australia • Brazil • Japan • Korea • Mexico • Singapore • Spain • United Kingdom • United States

Maths in the Real World - On the Building Site
1st Edition
Faye Cowin

Cover and text design: Cheryl Rowe
Typeset: Book Design Ltd, www.bookdesign.co.nz
Production controller: Siew Han Ong

Any URLs contained in this publication were checked for currency during the production process. Note, however, that the publisher cannot vouch for the ongoing currency of URLs.

First published in 1999 as Putting Maths to Work - On the Building Site by New House Publishing

Acknowledgements
Cover images courtesy of Shutterstock.
Images on pages 11, 14, 15, 23, 32, 37, 41, 45, 46, 50, 51, 52, 54, 56, 57, 58, 60, 61, 62, 64, 69, 70, 73, 75, 77, 78, 79, 80, 82, 83, 84, 90, courtesy of Shutterstock.
House plans on pages 34, 36, 38, 59 courtesy of Laing Homes.

For product information and technology assistance,
in Australia call **1300 790 853**;
in New Zealand call **0800 449 725**

For permission to use material from this text or product, please email **aust.permissions@cengage.com**

National Library of New Zealand Cataloguing-in-Publication Data
Cowin, Faye.
On the building site / Faye Cowin.
(Maths in the real world)
Previous ed: 1999, in series: Putting mathematics to work.
ISBN 978-017023-867-0
1. Building—Mathematics—Problems, exercises, etc. I. Title.
II. Series: Cowin, Faye. Maths in the real world.
690.0151—dc 23

Cengage Learning Australia
Level 7, 80 Dorcas Street
South Melbourne, Victoria Australia 3205

Cengage Learning New Zealand
Unit 4B Rosedale Office Park
331 Rosedale Road, Albany, North Shore 0632, NZ

For learning solutions, visit **cengage.com.au**

Printed in Singapore by 1010 Printing Group Limited
3 4 5 6 20 19

Maths
IN THE
Real World

On the
BUILDING
Site

Faye Cowin

NELSON
CENGAGE Learning

Australia • Brazil • Japan • Korea • Mexico • Singapore • Spain • United Kingdom • United States

Contents

Introduction 6

Literacy and Numeracy Standards for NCEA Level One 9

On the building site 10

1 Getting started 12

2 Subdivisions 15

3 Sections 23

4 House plans 32

5 Building a house 37

    Becoming a builder 37

    Becoming a roofer 41

    Becoming a bricklayer 46

6 Kitchens and bathrooms 50

    Becoming a plumber/gasfitter 50

    Becoming a drainlayer 51

    Becoming an electrician 54

7 Paint, paper and plaster 57

    Becoming a painter or decorator 57

8 Covering the floor 64

    Becoming a carpet/vinyl layer or tiler 64

9 Window dressing 70

10 Swimming pools and spas 73

11 Structures 77

    Bridges 77

    Kites 78

    Flying objects 79

    Tables and chairs 79

    Towers 80

    Letterboxes 82

    Stairs and steps 83

    Gates 84

12 Putting it into practice 86

    Achievement Standard 91026 86

    Achievement Standard 91030 88

    Achievement Standard 91032 89

Enrichment 90

Answers 92

Introduction

Maths in the Real World is a series of theme based books identifying the necessary mathematical skills and knowledge needed in particular areas for the future. The books cover key aspects of the *New Zealand Maths and Statistics for New Zealand curriculum*. The skill requirements of the NCEA Level One achievement standards and remaining unit standards are covered, enabling students to be examined in these NCEA qualifications. They also address the requirements of the Numeracy Project.

Maths in the Real World is aimed at those students who wish to pursue a non-academic career but for whom mathematics is an essential component of their trade training or life in the future. The series is an alternative course in mathematics for 15 to 17 year-olds in schools or training centres. Together, the books offer students:

- Practice with basic mathematics and calculation skills, so essential in the transition from school to the next stage in their development – flatting, travelling, working, buying a car.
- Opportunities to apply mathematics in practical everyday situations – making budgets, shopping, earning money, paying bills, planning a trip, owning a car.
- An awareness of their individual rights and responsibilities and an introduction to the range of community facilities available to them.

Essential skills covered in this book

- Whole number and all operations
- Problem solving
- Rounding of decimals
- Decimals and all operations
- Form filling
- Percentages.

ISBN: 9780170238670

New Zealand curriculum level 5: Numbers and algebra

Number strategies and knowledge

- Reason with linear proportions.
- Use prime numbers, common factors and multiples and powers (including square roots).
- Understand operations on fractions, decimals, percentages and integers.
- Use rates and ratios.
- Know commonly used fraction, decimal and percentage conversions.
- Know and apply standard form, significant figures, rounding and decimal place value.

Measurement

- Select and use appropriate metric units for length, area, volume and capacity, weight (mass), temperature, angle and time, with awareness that measurements are approximate.
- Convert between metric units using decimals.
- Deduce and use formulae to find the perimeters and areas of polygons and the volumes of prisms.
- Find the perimeters and areas of circles and composite shapes and the volumes of prisms including cylinders.

Statistics

Statistical investigation

- Plan and conduct surveys and experiments using the statistical enquiry cycle:
    - determining appropriate variables and measures
    - considering sources of variation
    - gathering and cleaning data
    - using multiple displays, and recategorising data to find patterns, variations, relationships and trends in multivariate data sets.
- Comparing sample distributions visually, using measures of centre, spread and proportion.
- Presenting a report of findings.

Statistical literacy

- Evaluate statistical investigations or probability activities undertaken by others, including:
    - data collection
    - methods
    - choice of measures
    - validity of findings.

ISBN: 9780170238670

New Zealand curriculum level 6: Numbers and algebra

Number strategies and knowledge

- Apply direct and inverse relationships with linear proportions.
- Extend powers to include integers and fractions.
- Apply everyday compounding rates.
- Find optimal solutions, using numerical solutions.

Equations and expressions

- Form and solve linear equations and inequations, quadratic and simple exponential equations and simultaneous equations with two unknowns.

Geometry and measurement

Measurement

- Measure at a level of precision appropriate to the task.
- Apply the relationships between units in the metric system, including the units for measuring different attributes and derived measures.
- Calculate volumes, including prisms, pyramids, cones, and spheres, using formulae.

Shape

- Deduce and apply the angle properties related to circles.
- Recognise when shapes are similar and use proportional reasoning to find an unknown length.
- Use trigonometric ratios and Pythagoras, theorem in two and three dimensions.

Position and orientation

- Use a coordinate plane or map to show points in common and areas contained by two or three loci.

Transformation

- Compare and apply single and multiple transformations.
- Analyse symmetrical patterns by the transformations used to create them.

ISBN 9780170238670

Literacy and Numeracy Standards for NCEA Level One

NCEA Level One Numeracy Standard	Number	Credits	New Zealand Curriculum
Apply numeric reasoning when solving problems	91026	4	NA6.1: Apply direct and inverse relationships with linear proportions NA6.2: Extend powers to include integers and fractions NA6.3: Apply everyday compounding rates
Solve measurement problems	91030	3	GM6.2: Apply the relationships between units in the metric system, including the units for measuring different attributes and derived measures GM6.3: Calculate volumes, including prisms, pyramids, cones and spheres, using formulae
Solve measurement problems involving right angled triangles	91032	3	GM6.1: Measure at a level of precision appropriate to the task GM6.5: Recognise when shapes are similar and use proportional reasoning to find an unknown length GM6.6: Use trigonometric ratios and Pythagoras' theorem in two and three dimensions
Apply transformation geometry	91034	2	GM6.8: Compare and apply single and multiple transformations GM6.9: Analyse symmetrical patterns by the transformations used to create them
Use the statistical enquiry cycle to investigate bivariate numerical data	91036	3	S6.1: Plan and conduct investigations using the statistical enquiry cycle

ISBN: 9780170238670

On the building site

Building a house or a shed or doing alterations (renovations) involves many mathematical skills, and there are several different types of tradespeople involved in the building industry. What are some of these trades? What are some of the mathematical skills they use?

What is perimeter?

What is area?

What is a regular shape?

What is an irregular shape?

ISBN: 9780170238670

Mathematical basics

Below are some of the formulae and measurements you will use in this book.

How do you measure up to this list?	
Area of a square/rectangle	$= L \times B$
Area of a triangle	$= \frac{1}{2} b \times h$
Area of a trapezium	$= \left(\frac{a+b}{2}\right) \times h$
Area of a parallelogram	$= h \times b$
Area of a circle	$= \pi r^2$
Perimeter is the distance around the outside of a shape	
Perimeter of a circle	$= \pi d$ or $2\pi r$
Pythagoras:	$= a^2 + b^2 = c^2$
Area of a non-right angled triangle	$= \frac{1}{2} ab \sin C$
Trigonometry ratios:	$\sin \Theta = \dfrac{\text{opposite side}}{\text{hypotenuse}}$
	$\cos \Theta = \dfrac{\text{adjacent side}}{\text{hypotenuse}}$
	$\tan \Theta = \dfrac{\text{opposite side}}{\text{adjacent side}}$
10 mm	= 1 cm
100 cm	= 1 m
1000 m	= 1 km
10 000 m²	= 1 ha
1 m³	= 1000 litres
1 cm³	= 1 mL
Volume	= area of base x ht. (units³)
Capacity	= quantity of fluid it can contain
Volume of a rectangular prism	$= l \times b \times h$
Volume of a cylinder	$= \pi r^2 h$
Ohm's Law: $\quad V = I \times R$	$F = C + 32 \times \frac{9}{5}$ $\quad$ $C = F - 32 \times \frac{5}{9}$

ISBN: 9780170238670

Getting started

ISBN: 9780170238670

Exercise 1

1 Convert the following fractions to decimals and percentages.

   a ½   **b** ⅗   **c** ⅞

   d ⅓   **e** 4⅜

2 Convert the following decimals to fractions and percentages.

   a 1.75   **b** 0.05   **c** 0.4

   d 2.55   **e** 9.24

3 Convert the following percentages to fractions and decimals.

   a 35%   **b** 250%   **c** 12.5%

   d 15%   **e** 5%

4 Use your calculator to find the squares of these numbers.

   a 6 mm   **b**   5.5 m

   c 0.35 m   **d**   2500 mm (answer m²)

5 Use your calculator to find the square root of these numbers.
(Answer to 3 decimal places.)

   a 625 mm   **b** 156.25 m

   c 48 m   **d** 1200 mm

6 Convert the following lengths to the units in brackets.

   a 1 m (mm)   **b** 5.25 m (mm)   **c** 2300 mm (m)

   d 56 cm (mm)   **e** 1550 mm (m)   **f** 40 mm (m)

   g 6.8 m (mm)   **h** 4.05 m (mm)

   i 165 cm (m)   **j** 0.56 m (mm)

7 Convert the following weights to the units in brackets.

   a 1000 g (kg)   **b** 2.5 kg (g)   **c** 1 tonne (kg)   **d** 40 kg (g)

8 Convert the following capacities to the units in brackets.

   a 1 L (mL)   **b** 0.25 L (mL)   **c** 2500 mL (L)   **d** 15 L (kL)

9 Find the perimeter and area of the following shapes.

   a 4 m by 4 m square   **b** 2500 mm by 1500 mm rectangle

   c   **d**

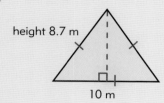

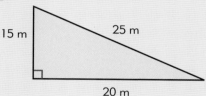

10 Find the volume of the following shapes.

   a 300 mm cube   **b** 500 mm by 250 mm by 700 mm

   c   **d**

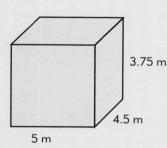

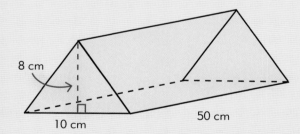

11 Find the circumference and volume of these cylinders. (Use π = 3.14.)

   a A cylinder has a radius of 450 mm and a depth of 1500 mm.
      (Volume answer in mm^3 and m^3.)

   b A pipe is 10 m long and it has a diameter of 100 mm.
      (Volume answer in mm^3 and m^3.)

   c A hot water cylinder has a radius of 280 mm and a height of 950 mm.
      (Volume answer in mm^3 and m^3.)

12 What is the capacity in litres of each of the cylinders in question 11 above.

ISBN: 9780170238670

13 Generally, concrete is mixed in the following ratios: two parts cement : one part water : ten parts builders mix (i.e. 2 : 1 : 10).

a If you want to make 13 m³ of concrete, how much of each do you need?

b If you want to make 52 m³ of concrete, how much of each do you need?

c If you have a 40 kg bag of cement, how much concrete will this make?

14 Below are some of the conversions between metric and imperial units:

2.2 lb = 1 kg
1 mile = 1.6 km
1 yard = 39 cm
1 inch = 2.54 cm
1 gallon = 3.8 litres

Using these conversions, calculate the following.
a 5 kg to lbs
b 28.5 L to gallons
c 2 miles to km
d 12 inches to cm

15 A US magazine says a 10-inch diameter pipe is needed to carry storm water. How big is this diameter in cm?

16 A box of special imported brass bolts from the US weighs 19.8 pounds (19.8 lbs). What is the weight in kilograms?

17 Convert these temperatures to °C.
a 50 °F
b 212 °F
c 0 °F
d 165 °F

ISBN: 9780170238670

Subdivisions

- What do 'industrial', 'commercial' and 'residential' areas mean?
- Are there different types within these categories?
- What are your local council regulations for building?
- What services do subdivisions need?
- How are sections labelled?

Exercise 2

Study the subdivision map on the following page and answer these questions.

1   How many sections are for sale in the subdivision?

2   Which section is greater than 900 m^2? Can you give a reason for its size?

3   Name the two sections on the corner of Chiswick Avenue and Sterling Gate Drive.

4   What are the dimensions of Lot 56?

5   Lot 30 is $170 000 and Lot 39 is $236 000. Write an explanation for the difference in price.

6   Jane is a postie in this subdivision. Sketch the subdivision and an economical route she could take starting at Lot 26.

7   What is the total area of all the sections? (Answer in ha.)

8   What is the average section size of all the sections? (Answer in m^2 and ha and to 3 significant figures.)

ISBN: 9780170238670

9 The area of all three streets in this subdivision is 1.1807 ha. Using your answer in question 7 above, what is the ratio of street area : total section area? (Answer to 1 significant figure.)

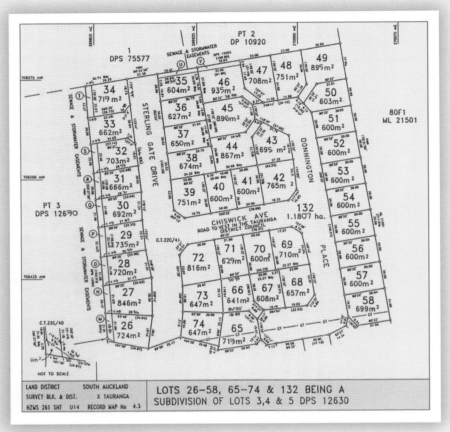

Exercise 3

In the subdivision plan on the following page there are ten sections. Study the plan and the table (below) to answer questions 1–4 on page 18.

Unit	BR	Garage	Section	Entrance	House size	Price
1	2	Single	271 m²	No	104 m²	$389 000
2	3	Single	293 m²	No	126 m²	$398 000
3	3	Double	384 m²	Yes	144 m²	$455 000
4	3	Single	317 m²	No	126 m²	$398 000
5	3	Single	294 m²	Yes	129 m²	$399 000
6	3	Single	347 m²	Yes	125 m²	$399 000
7	3	Double	290 m²	Yes	131 m²	$409 000
8	3	Double	296 m²	Yes	131 m²	$415 000
9	3	Double	305 m²	No	127 m²	$428 000
10	3	Double	313 m²	No	124 m²	$429 000

ISBN: 9780170238670

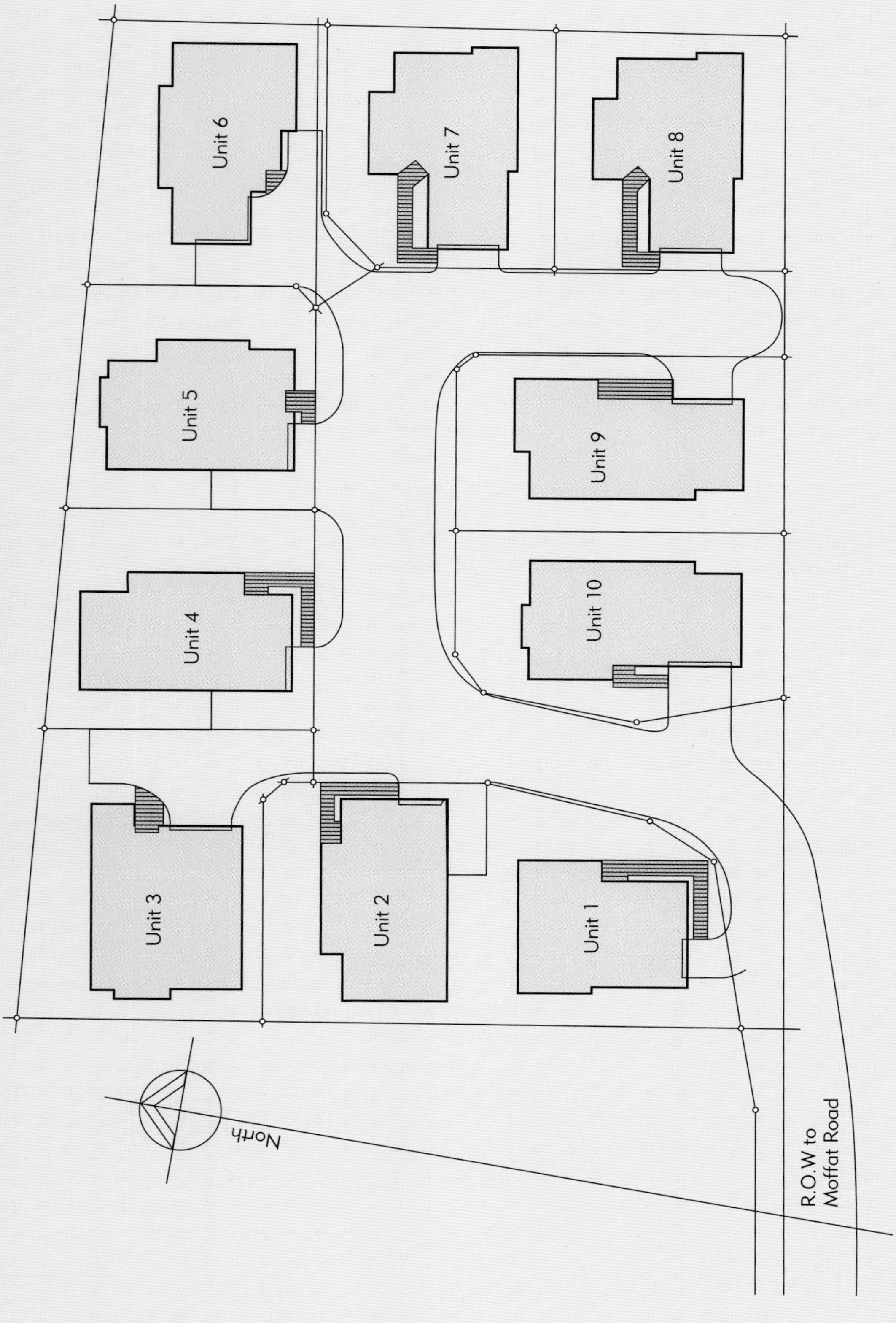

Unit 6

Unit 7

Unit 8

Unit 5

Unit 9

Unit 4

Unit 10

Unit 3

Unit 2

Unit 1

North

R.O.W to
Moffat Road

ISBN: 9780170238670

1   What is the median house and section price?

2   What percentage (answers to 1 decimal place) of the section size is the house size in …

   a   Lot 1

   b   Lot 6

   c   Lot 7

3   **a**   What is the cost of buying …

      i   Unit 3

      ii   Unit 10

   b   Can you give an explanation for Unit 3's price?

4   The prices are for the section and the house. Deduct $105 000 for each section and calculate the average price per m² of the houses.

   a   Unit 4

   b   Unit 1

   c   Unit 10

5   Study these tables and answer the questions below.

Sections			
Woodbury Rise	Lot 120	659 m²	$157 000
	Lot 123	767 m²	$154 000
	Lot 119	688 m²	$158 000
	Lot 152	600 m²	$168 000
Cypress Grove (min. house size 110 m²)	Lot 8	664 m²	$173 000
	Lot 39	663 m²	$171 000

Houses			
Sommerset	81.3 m²	3 BR – Hardiplank – Brick	$339 500 $378 500
Resolution 3	99.6 m²	3 BR – Hardiplank – Brick	$358 500 $390 500
Resolution 4	120.6 m²	4 BR – Hardiplank – Brick	$397 000 $399 900
Endeavour 3	102.5 m²	3 BR – Hardiplank – Brick	$375 000 $392 000
Endeavour 4	122.6 m²	4 BR – Hardiplank – Brick	$405 300 $423 800
Discovery	135.3 m²	4 BR – Brick	$435 100

ISBN: 9780170238670

   a What would it cost per square metre to buy Lot 119?

   b Which section costs the most per m²?

   c What would it cost per square metre to have these houses built?

      i 3 BR Endeavour with hardiplank cladding

      ii 4 BR Discovery

   d How much extra does it cost to have Endeavour 4 built in brick cladding, rather than hardiplank?

   e Sue and James decide to buy Lot 8 and a 4 BR brick Endeavour house.

      i What is the total cost?

      ii What percentage is the house cost of the total cost?

   f What houses could you put on Lot 39? Why do you think this is stated?

6   If real estate agents charge 3.95% on the first \$400 000, then 2.5% thereafter, what is the fee charged on these sales?

   a \$280 000

   b \$458 000

   c \$695 000

   d \$155 000

   e Lot 8 and the 4 BR brick Endeavour in question 5e?

7   The table below shows the New Zealand average house prices from 2006 to 2011 (Data from the *NZ Real Estate Guide*).

	Average House Price (May 2011)	Average House Price (May 2010)	Average House Price (May 2009)	Average House Price (May 2008)	Average House Price (May 2007)	Average House Price (May 2006)
Auckland	\$526 000	\$517 000	\$483 000	\$514 000	\$488 000	\$450 000
Hamilton	\$330 000	\$330 000	\$324 000	\$348 000	\$318 000	\$289 000
Tauranga	\$330 000	\$354 000	\$340 000	\$368 000	\$355 000	\$330 000
Wellington	\$446 000	\$457 000	\$412 000	\$421 000	\$416 000	\$361 000
Christchurch	\$402 000	\$410 000	\$377 000	\$417 000	\$398 000	\$356 000
Nelson	\$335 000	\$337 000	\$320 000	\$330 000	\$312 000	\$280 000
Queenstown	\$500 000	\$510 000	\$510 000	\$550 000	\$520 000	\$516 000
Dunedin	\$245 000	\$249 000	\$239 000	\$250 000	\$245 000	\$230 000

ISBN: 9780170238670

a   Which district has the highest average median over the six years?
b   Can you give some reasons for your answer in a above?
c   Which district has the lowest average median over the six years?
d   Can you give some reasons for your answer in c above?
e   Which district had the biggest percentage increase in value from 2006 to 2011?
f   Draw a line graph using the data from the table above to show the average house prices over the six years for each of the districts.
g   Write an analysis and conclusion about your graph, covering such aspects as highs, lows, clusters and trends.

8   This table shows the median house prices from 2006 to 2011 for seven areas of Auckland.

Location	Average House Price (May 2011)	Average House Price (May 2010)	Average House Price (May 2009)	Average House Price (May 2008)	Average House Price (May 2007)	Average House Price (May 2006)
Auckland North – Albany	$640 000	$620 000	$620 000	$630 000	$590 000	$523 000
Auckland North – Birkenhead	$520 000	$535 000	$483 000	$515 000	$490 250	$458 000
Auckland Central – Eastern Suburbs	$800 000	$770 000	$720 000	$765 000	$730 500	$660 000
Auckland Central City	$315 000	$350 000	$325 000	$349 000	$356 250	$345 000
Auckland Central – Mt Eden / Epsom	$690 000	$652 000	$551 000	$630 000	$577 000	$556 000
Auckland West – Henderson Area	$370 000	$365 000	$357 000	$370 000	$350 000	$328 000
Auckland South – Papatoetoe	$345 000	$330 000	$325 000	$340 000	$320 000	$283 000

ISBN: 9780170238670

a Over the past six years, which area is the most expensive to live in?

b Over the past six years, which area is the least expensive to live in?

c Can you give a reason for the prices in 2008?

d What is the percentage increase in prices in 2006 and 2011 for the Henderson area?

e Draw a line graph of all the above data and write an analysis of what you notice.

9 This table shows the median house prices from 2006 to 2011 for the five areas of Christchurch.

ISBN: 9780170238670

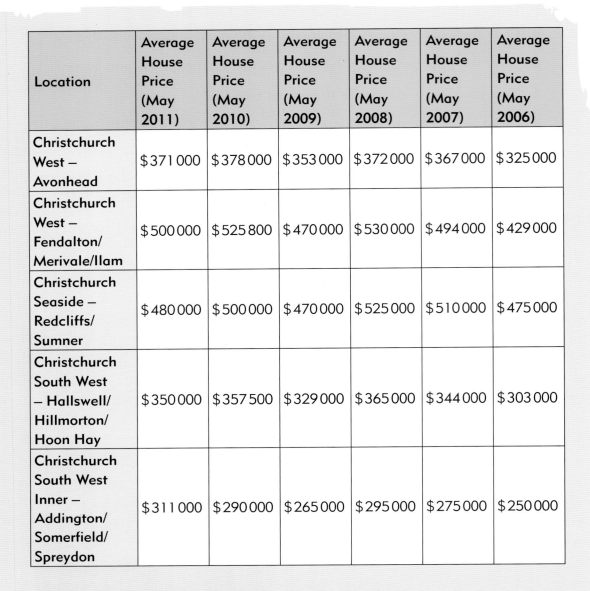

Location	Average House Price (May 2011)	Average House Price (May 2010)	Average House Price (May 2009)	Average House Price (May 2008)	Average House Price (May 2007)	Average House Price (May 2006)
Christchurch West – Avonhead	$371 000	$378 000	$353 000	$372 000	$367 000	$325 000
Christchurch West – Fendalton/ Merivale/Ilam	$500 000	$525 800	$470 000	$530 000	$494 000	$429 000
Christchurch Seaside – Redcliffs/ Sumner	$480 000	$500 000	$470 000	$525 000	$510 000	$475 000
Christchurch South West – Hallswell/ Hillmorton/ Hoon Hay	$350 000	$357 500	$329 000	$365 000	$344 000	$303 000
Christchurch South West Inner – Addington/ Somerfield/ Spreydon	$311 000	$290 000	$265 000	$295 000	$275 000	$250 000

a What was the most expensive area to live in 2007?

b What was the percentage increase in price for Avonhead in 2006 and 2011?

c What is the average house price for Christchurch for 2008?

d Over the past six years, which area has had the greatest percentage increase in price?

e Can you give a reason(s) why Christchurch South West Inner houses may have increased in price?

ISBN: 9780170238670

3

Sections

- What is the size of the land you live on?
- What is the size of an average house section?
- What is a unit site?
- Does your local council have regulations on minimum size for a dwelling?
- What is Residential A and B zoning?

Exercise 4

1   Find the area of these sections. You may need to refer to page 11 to find some useful formulae. (Answer in m².)

a

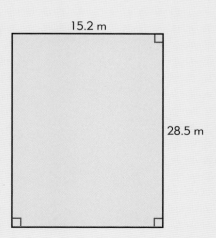

15.2 m

28.5 m

ISBN: 9780170238670

b

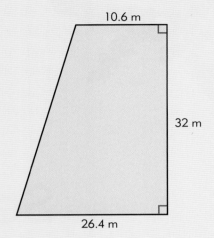

10.6 m

32 m

26.4 m

c

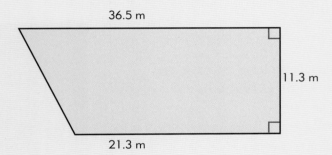

36.5 m

11.3 m

21.3 m

d

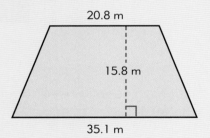

20.8 m

15.8 m

35.1 m

e

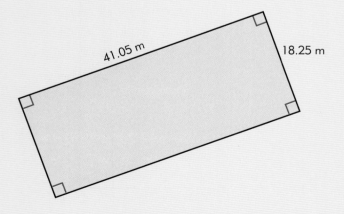

41.05 m

18.25 m

ISBN: 9780170238670

2   Not all sections are these regular shapes; in fact more sections are an irregular shape, like the following. To find the area of these shapes, each shape needs to be divided into regular shapes that can be calculated. Study these two examples then do the problems that follow.

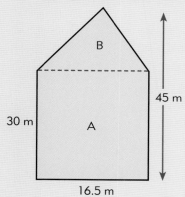

Area of A = 16.5 m x 30 m   = 495 m²

Area of B = ½ x 16.5 m x 15 m   = 123.75 m²

Total area = 495 m² + 123.75 m² = 618.75 m²

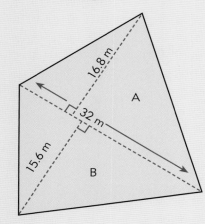

Area of A = ½ x 32 m x 16.8 m   = 268.8 m²

Area of B = ½ x 32 m x 15.6 m   = 249.6 m²

Total area = 268.8 m² + 249.6 m² = 518.4 m²

a

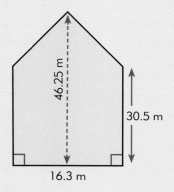

b

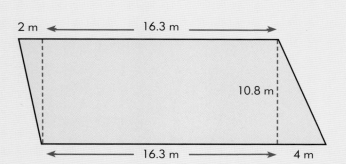

ISBN: 9780170238670

c

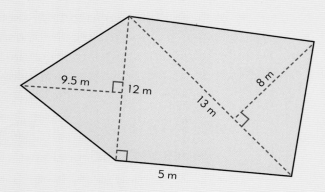

3 Find the perimeter of these sections.

a 30 m x 20 m rectangle.

b

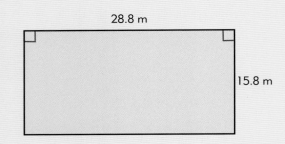

c

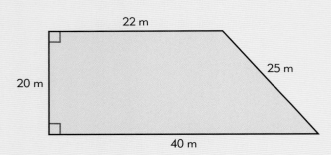

4 Find the area of each of the following fields.

a

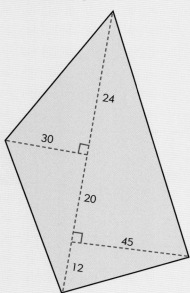

ISBN: 9780170238670

b

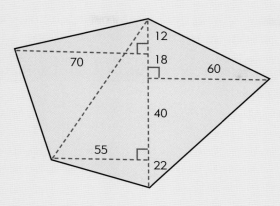

c

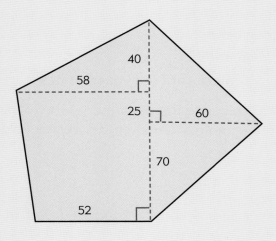

d

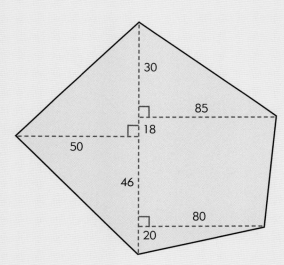

5 A rugby field is 100 m long and 65 m wide. How many complete fields could be placed on a square hectare?

ISBN: 9780170238670

6 **a** Copy this table into your book and investigate the number of ways a hectare can be formed from a rectangular shape.

No.	Dimensions	Area	Perimeter
1	100 m x 100 m		
2			

b Which dimensions give the …

i greatest perimeter?

ii smallest perimeter?

iii Write a few sentences about your findings.

7 Use Pythagoras to calculate the perimeter of the following sections. Measurements are in metres. (Answer to 1 decimal place.)

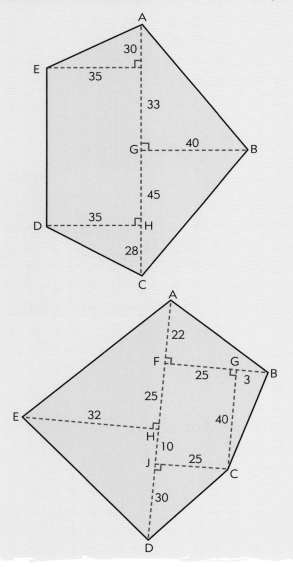

8 Using a scale 1 : 100 and, where necessary, a protractor, draw the following sections to scale.

a

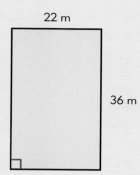

22 m

36 m

b

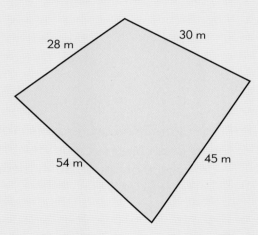

28 m

30 m

54 m

45 m

c

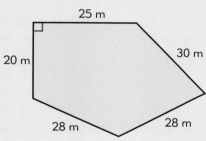

25 m

20 m

30 m

28 m

28 m

9 a What is the area of the section?

b What is the area of the house?

c What percentage of the section is covered by the house? (Answer to 1 decimal place.)

d What area of the section is uncovered?

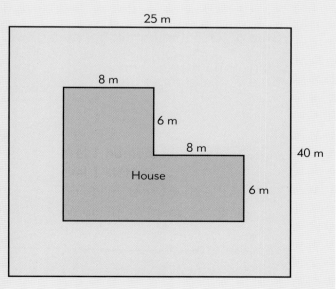

25 m

8 m

6 m

8 m

40 m

House

6 m

ISBN: 9780170238670

10 Lot 2135 (ABCD) has an area of 840 m².

  a What is the floor area of the house?

  b What area of the section is uncovered?

  c What is the area in a as a fraction of the whole section? (Simplify)

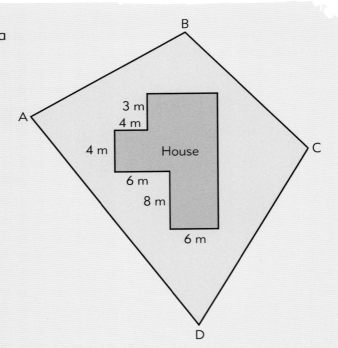

11 Land is surveyed to calculate its measurements. You can make your own plane table radial survey of an area such as your playing fields. You will need a plane table, large piece of paper, 30 m measuring tape, protractor and a straight edge with sights (alidade). Refer to the diagram and follow the steps below.

Step 1 Put four markers on the outer limits of the field to form an irregular quadrilateral.

Step 2 Put the plane table somewhere near the middle of the quadrilateral.

Step 3 Look through the alidade to each of the four markers, writing a line on the paper to indicate this direction.

Step 4 Use the tape to measure the distance from the table to each of the four markers and note this distance on the paper.

Step 5 Measure the angle between each distance (marked A).

Step 6 Use the formula for area ($A = \frac{1}{2}\,ab\sin C$) to calculate the area of each triangle.

Step 7 Total these areas.

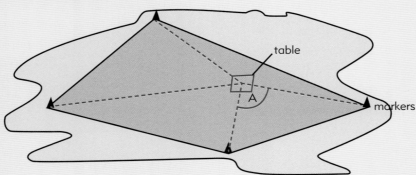

ISBN: 9780170238670

Now use this method to calculate the area of these fields.
(Answers to 1 decimal place.)

a

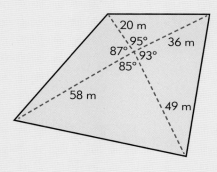

b

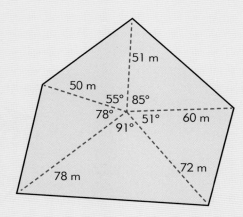

12 This plan is drawn to a scale of 1 : 500. Find by measurement …

   a   the dimensions of the section.

   b   the area of the section.

   c   the distance the house is from the northern boundary.

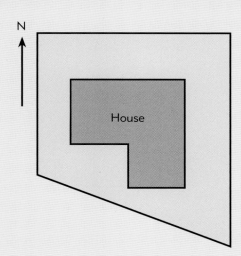

ISBN: 9780170238670

House plans

- What occupations use scale drawings?
- What do the scales mean?
- What is an elevation?
- What unit of length measurement is a house usually drawn in?

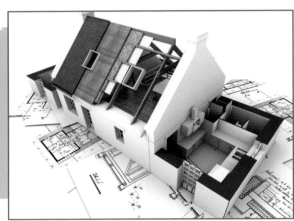

Exercise 5

1 What do these symbols represent?

a WC

b

c WD

d

e

f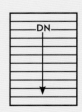

2 If a measurement on a plan is 50 mm, what lengths are these scales?

a 1 : 10

b 1 : 50

c 1 : 100

d 1 : 250

e 1 mm : 25 cm

f 1 mm : 12 cm

ISBN: 9780170238670

3 If a plan has a scale of 1 : 150, what lengths would these measurements represent?

   a 8 mm   **b** 30 mm   **c** 12 mm

   d 46 mm   **e** 27 mm   **f** 54 mm

4 Using this plan of a garage and workshop, measure the lengths with a ruler to show that the scale is 1 : 100.

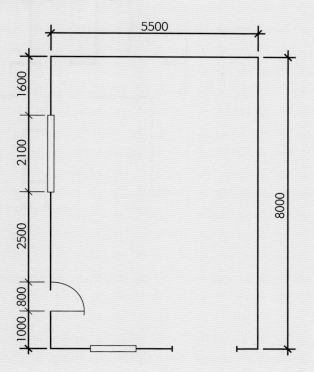

5 Using the plan of a bathroom below, what is the …

   a real length of x?

   b real length of y?

   c real area of the bathroom?

   d area of the shower? (Estimate)

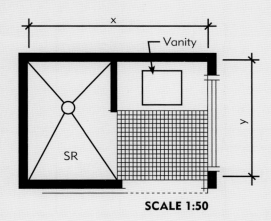

SCALE 1:50

ISBN: 9780170238670

6 **a** Find the real length of lengths marked a, b, c, d, e and f in the following plan.

  b What is f?

  c What is the approximate area of the living, dining and kitchen area?

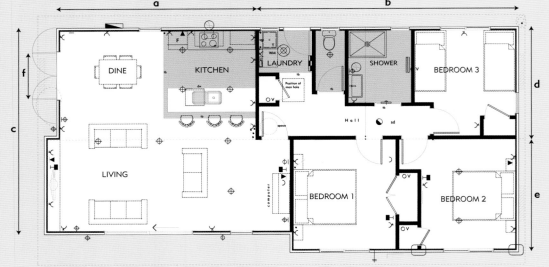

7 **a** What is the total area of the house below, excluding deck?

  b Estimate the area of carpet needed for the three bedrooms, lounge and dining areas.

  c What is the approximate area of the suggested deck area?

  d What percentage of the living area (lounge, dining, kitchen) would the deck be?

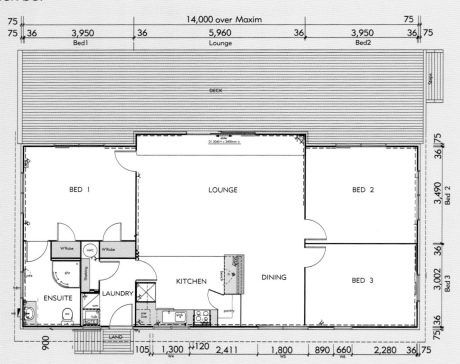

ISBN: 9780170238670

Plans are also drawn with elevation views. These views are from the north, south, west and east and enable you to visualise what the building will look like from these directions.

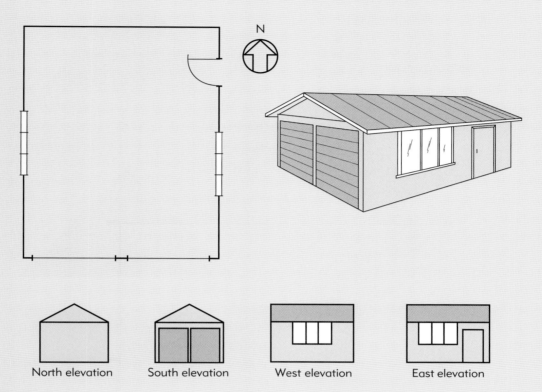

North elevation   South elevation   West elevation   East elevation

8 Sketch the four elevation views of this garage.

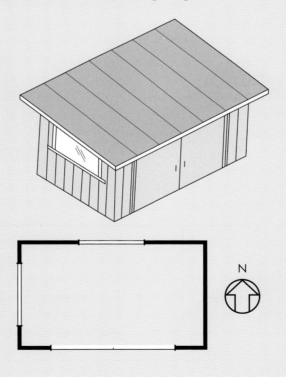

ISBN: 9780170238670

9 **a** What is the total area of this teaching facility, including classroom and lobby?

b What percentage is the resource room of the total area in question a? (Answer to 1 decimal place.)

c What are dimensions of the …

i classroom

ii resource room

iii lobby

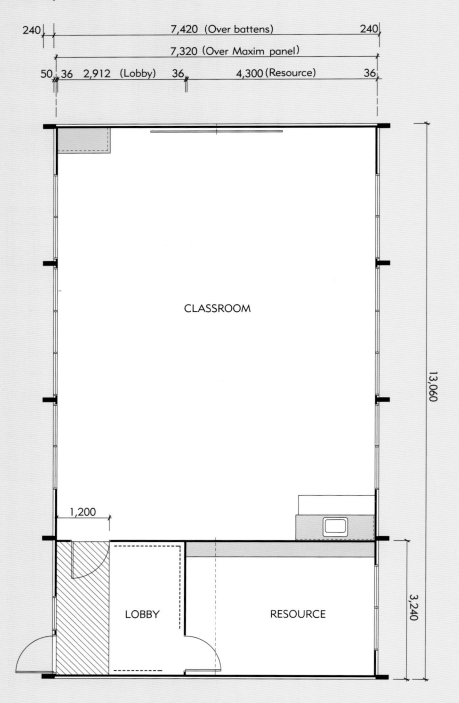

ISBN: 9780170238670

Building a house

- What is a permit?
- What do you need a permit for?
- Does your local council have any specific regulations?
- What is a registered tradesperson?
- Do you have to employ licensed practitioners?
- Why must all buildings have sound foundations?
- How many types of foundations can you think of?

Becoming a builder

Becoming a qualified builder enables you to work on any building site, whether it is residential or commercial, no matter how big or small the job may be. There is a wide variety of specialised trades within the building industry that you can pursue when you become qualified. The first qualification is the National Certificate in Construction, a Level 2 qualification that requires 74 credits. The Building and Construction Industry Training Organisation (BCITO) is the provider.

How do you get started?

You need to find an employer who is willing to take you on as an apprentice for the three to four years it will take. You will be given a training advisor who will support and mentor you during your apprenticeship. This will cover both your practical and theory work.

Skills you will need

- Level 1 numeracy.
- Level 1 literacy.
- The ability to work as a team.
- The ability to meet deadlines.
- A willingness to listen and learn.
- A satisfactory level of fitness and health.
- An awareness of health and safety procedures.
- A current driver's license is beneficial.

Apprenticeships are competency based, so learning on the job is an integral part of the process. You will need to regularly demonstrate your ability to complete tasks competently. Further details can be found at *www.bcito.org.nz* and the NZQA website *nzqa.govt.nz*.

'Footings' are trenches dug into the soil at a specified depth to support a concrete floor. The trench is in-filled with concrete and metal rods.

1 Copy this table into your book and calculate the volume of concrete required for these footings.

	Width	Depth	Length	Volume
a	300 mm	200 mm	1000 m	
b	400 mm	250 mm	4 m	
c	600 mm	300 mm	16.5 m	
d	500 mm	250 mm	31 m	
e	450 mm	220 mm	10 m	
f	550 mm	325 mm	8.5 m	

2 If concrete costs \$235.00 per m^3 (including GST), what would be the cost of concrete in each part of question 1?

3

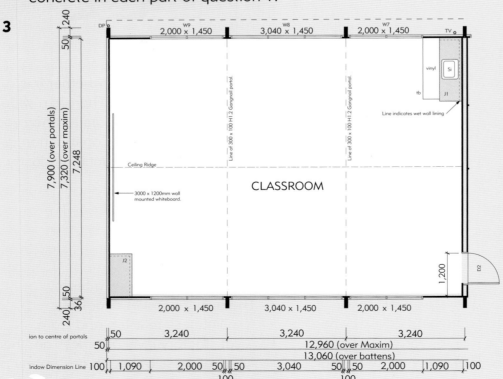

a Estimate how much concrete would be required for the footings for this classroom (page 38) if the footings are 200 mm wide and 300 mm deep.

b If the concrete cost $235.00 per m³ (including GST) delivered, how much would the concrete cost?

4 Once the footings are in place, concrete blocks are cemented on top of the footing to form a retaining wall for the sand, mesh and concrete. If concrete blocks are 400 mm long, how many would be required for these lengths?

a 4 m
b 25 m
c 86 m
d 73.5 m
e 102.5 m

5 Concrete floors are usually 100 mm thick. Copy this table into your book and calculate the volume and cost of concrete for each of these areas.

	Length	Width	Volume	Cost ($235/m³ inc. GST)
a	15 m	10 m		
b	18.5 m	12 m		
c	22.8 m	10.5 m		
d	16.4 m	15.5 m		
e	area = 253.5 m²			

6 Piles and poles are another form of foundation for a building. Piles must be placed no more than 500 mm apart around the perimeter of the building and at strategic positions inside the area. How many piles would be required for the perimeter of these floor areas if there has to be a pile at each corner and they must be no more than 500 mm apart?

a

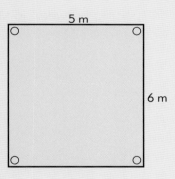

b

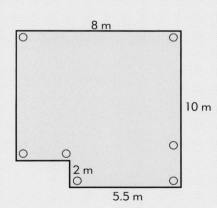

7 The diagram below shows individual, running and overall measurements.

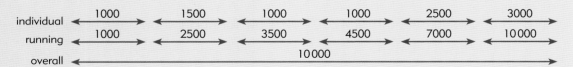

ISBN: 9780170238670

What are the missing measurements in the following diagrams?

a

| 1000 | 1200 | 1800 | 1000 | 1500 | 1300 |

i

ii

b i

| 1200 | 2500 | 5000 | 7500 | 9200 | 9500 |

ii

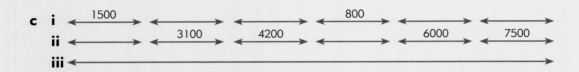

c i

| 1500 | | | 800 | | |

ii

| | 3100 | 4200 | | 6000 | 7500 |

iii

8 Framing members (floor joists or studs) need to be evenly spaced. How many floor joists at 600 mm centres would be needed for these lengths?

a 3000 mm
b 1800 mm
c 4800 mm

9 Spacings between studs should be no more than 600 mm. The distance between can be measured by:

$$\frac{\text{overall length} - \text{space taken up by all framing members}}{\text{number of spacings}}$$

The formula below shows how to calculate the spacing between joists. Each end stud is 80 mm wide and the studs in between are 40 mm.

$$\frac{2000 - (80 + 40 + 40 + 40 + 80)}{4}$$

$$= 430 \text{ mm centres}$$

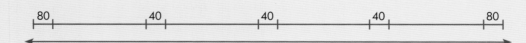

| 80 | 40 | 40 | 40 | 80 |

a If you have a length of 3200 mm, how far apart should each stud be?

b If you have a length of 4200 mm, how far apart should each stud be?

c If you have a length of 2650 mm, how far apart should each stud be?

ISBN: 9780170238670

Becoming a roofer

- Roofs can be of different styles and made of different materials. How many can you think of?
- How were roofs made before the twentieth century?
- Do roofs need to be insulated?

There are seven NZQA roofing National Certificates and all are monitored by the Roofing Industry Training Organisation (RITO). Roofing apprentices learn on the job with an employer engaged in the day to day supervision of their training. A metal roofing apprenticeship is usually in four stages over a period of approximately two years. At the end of each year apprentices will receive a record of their learning, listing the unit standards completed for that year.

How do you get started?

Like other trades in the building industry, you will need to find an employer who is willing to take you on as an apprentice as well as provide supervision for your apprenticeship time. Further detailed information can be found at *www.roofingassn.org.nz*

Skills you will need

- Level 1 numeracy.
- Level 1 literacy.
- Little or no fear of heights.
- The ability to work as a team.
- The ability to meet deadlines.
- A willingness to listen and learn.
- A satisfactory level of fitness.
- An awareness of health and safety issues.
- A current driver's license is beneficial.

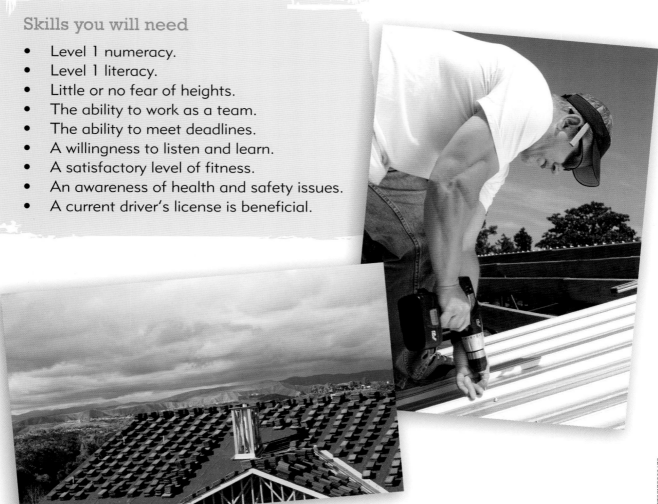

ISBN: 9780170238670

Exercise 7

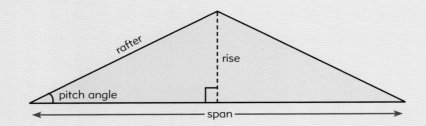

rafter

rise

pitch angle

span

1 Use Pythagoras to calculate the rafter length (one side) of each of these roofs. (Answer to 1 decimal place.)

a

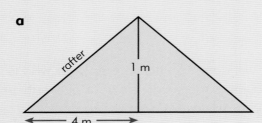

rafter

1 m

4 m

b

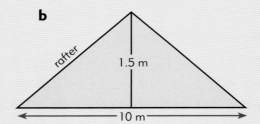

rafter

1.5 m

10 m

c

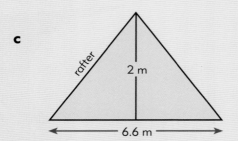

rafter

2 m

6.6 m

d

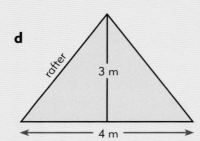

rafter

3 m

4 m

2 Calculate the height of the roof (x). (Answer to 1 decimal place.)

a

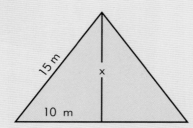

15 m

x

10 m

b

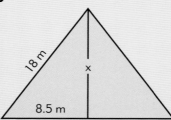

18 m

x

8.5 m

ISBN: 9780170238670

3 Calculate the span of the roof (y). (Answer to 1 decimal place.)

a

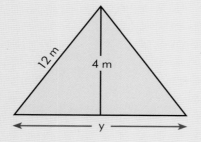

b

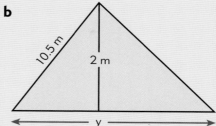

c

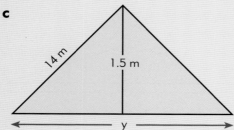

d

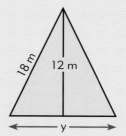

4 Calculate the length of the hip run (X) in each of the following diagrams. (Answer to 1 decimal place.)

a

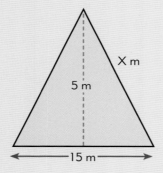

b

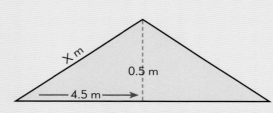

c

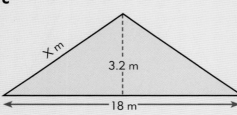

d

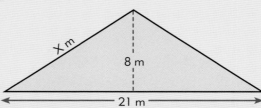

ISBN: 9780170238670

5 Use the trigonometry ratios to calculate the angle of the roof pitch (x).

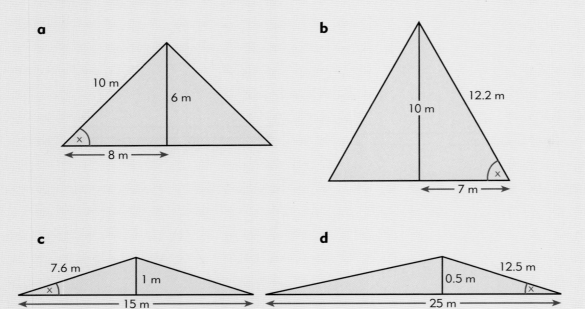

a

10 m   6 m

8 m

b

12.2 m

10 m

7 m

c

7.6 m   1 m

x

15 m

d

12.5 m

0.5 m

x

25 m

6 Roofing iron is usually 760 mm wide and is cut to the length of the rafter. How many metres of roofing iron would be required to cover these roofs? Note that overlap is included in the 760 mm. (Answer to nearest metre.)

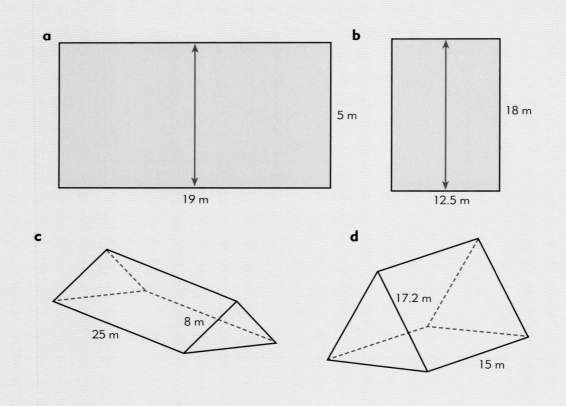

a

5 m

19 m

b

18 m

12.5 m

c

25 m   8 m

d

17.2 m

15 m

ISBN: 9780170238670

7 This shows a roof section drawn to scale 1 : 50.

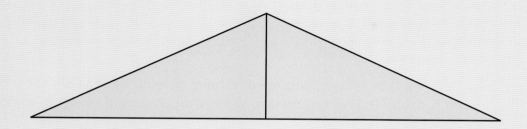

a Calculate the rafter length.

b Calculate the rise.

c Calculate the span length.

d Measure the pitch angle.

e If the building was 25 m long, how many metres of roofing iron would be needed?

f If the roofing iron costs $44.75 per m², what is the cost of the iron?

8 Roofing tiles are usually 420 mm long and 330 mm wide.

a What is the area of one tile in mm² and m²?

b How many tiles would be needed to cover these areas?

   i   6 m × 10.5 m

   ii   12.5 m × 8.2 m

   iii   11.6 m square

   iv   105 m²

   iv   185.5 m²

ISBN: 9780170238670

Becoming a bricklayer

- Exterior walls can be covered in a variety of claddings of various costs.
- What is the most popular in your area?
- Is there a reason for this popularity?

If you want to become a bricklayer, you need to have at least three years secondary schooling, and it is useful (but not essential) to have studied mathematics, woodwork and technical drawing.

How do you get started?

Like the builder, you need to find an employer who is willing to employ you while you train. A qualified bricklayer will oversee your practical and theory work. You need to complete all the required unit standards and 8000 hours of practical work.

Skills you will need

- Level 1 numeracy.
- Level 1 literacy.
- The ability to work as a team.
- The ability to meet deadlines.
- A willingness to listen and learn.
- A satisfactory level of fitness and health.
- An awareness of health and safety procedures.
- A current driver's license is beneficial.

Exercise 8

1
- Bricks are manufactured to New Zealand nominal size 230 mm x 76 mm x 90 mm
- 48.6 bricks cover 1 m^2
- 968 bricks per pallet
- Average price $2.75 each

How many bricks would be needed to cover …

a 50 m^2

b 16 m^2

c 65 m^2

d 3.5 m^2

e this rectangular-shaped house (Ignore windows and doors.)

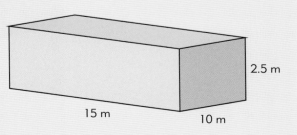

15 m · 10 m · 2.5 m

ISBN: 9780170238670

2 This is the New Zealand nominal size of a brick. Using these measurements, what area would the following number of bricks cover? (Answer in m², and to 2 decimal places for questions b–f.)

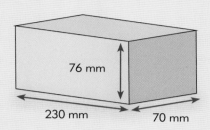

76 mm

230 mm

70 mm

a	1	**b**	10	**c**	350
d	1240	**e**	3680	**f**	2973

3 What area does one pallet of bricks cover? (Answer to 2 significant figures.)

4 How many pallets of bricks would you need to cover these areas?

a 50 m²   **b** 72 m x 2.5 m   **c** 84 m x 3 m

5 Weatherboards, ranchsliders and windows are usually installed by the builder, but sometimes specialised tradespersons will complete the task. The following table shows nominal dimensions of a range of weatherboards.

Weatherboard dimensions				
Type	Length (mm)	Width (mm)	Effective cover (mm)	Thickness (mm)
Rusticated	4200	205	175	7.5
Styleline	4200	205	175	7.5
Smooth	4200	180	150	7.5
	4200	240	210	7.5
Colonial	4200	205	175	7.5
Summit	4200	150	120	9
Frontier	4200	245	215	7.5
	4200	310	280	7.5

a What is meant by effective cover?

b How many millimetres does one Styleline weatherboard cover using …
   i One width?
   ii Effective cover?
   iii What percentage of the weatherboard is overlapped?

ISBN: 9780170238670

c John wanted to put rusticated weatherboards on this wall.

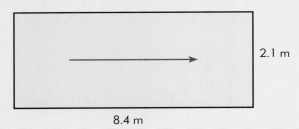

2.1 m

8.4 m

    i   What is the effective width cover of the weatherboard?

   ii How many widths would be needed to cover 2.1 m?

  iii If each plank is 4200 mm long, what is the total length of weatherboards he would need? (Give your answer in metres.)

  iv If this weatherboard costs \$85 per m², how much will it cost?

6 James wants to install bandsawn Cedar weatherboards which come in 4200 mm lengths, with effective cover of 175 mm, and an actual width of 205 mm.

  a How many lineal metres of the Cedar weatherboard would be needed to cover this area?

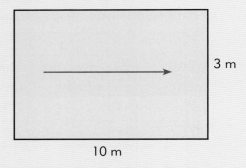

3 m

10 m

  b If the weatherboard costs \$189 per m² how much will it cost?

7 Make a sound estimate of how many lineal metres of the wide Smooth weatherboard is required to cover the following wall plan.

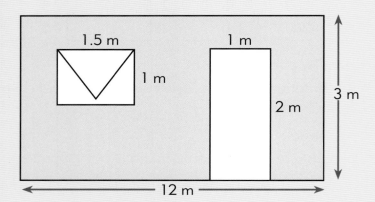

1.5 m

1 m

1 m

2 m

3 m

12 m

ISBN: 9780170238670

8 On a plan, windows (top) and ranchsliders (bottom) are drawn like this:

a How many ranchsliders and windows are there altogether in this house plan?

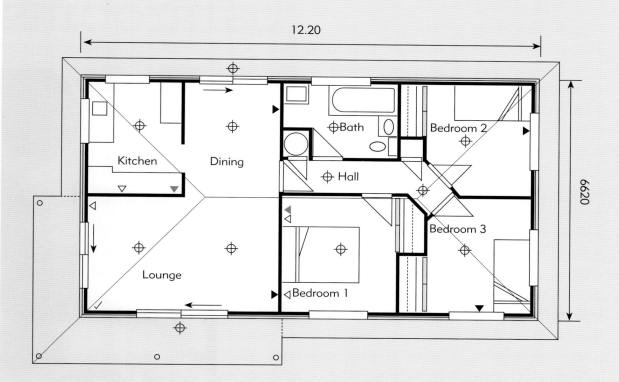

b If the scale is 1 : 100, what is the length of the …
  i two lounge ranchsliders?
  ii bedroom 1 window?
  iii kitchen window?
c If ranchsliders are 2 m in height and the windows average 1 m, what area of the walls is covered by glass?

ISBN: 9780170238670

Kitchens and bathrooms

Becoming a plumber/gasfitter

Plumbing and gas fitting careers offer great rewards, and the skills and knowledge learned are transferable to many other sought after areas in the construction industry.

How do you get started?

Like a builder, you need to find an employer willing to employ you as an apprentice. Plumbing and gasfitting is a complex trade requiring lots of knowledge and skills, so you do need good literacy and numeracy skills.

The apprenticeship takes four years and involves work on the job as well as distance learning and block courses at polytechnics. The length of this apprenticeship is indicative of the health and safety issues involved. These must be implemented in a

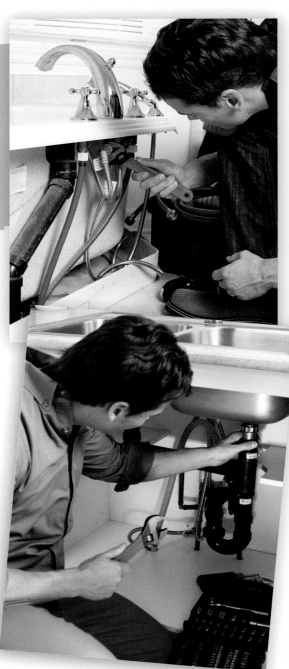

ISBN: 9780170238670

safe and professional manner. Once you have completed all of these requirements, you sit an examination to become licenced. As an apprentice you are given a limited licence, but once you have successfully completed your apprenticeship you become a licenced plumber and gasfitter. Your work still needs to be supervised for a further two years to ensure it is safe and completed properly. After this time you become a certified plumber and gasfitter. More information can be found at *www.ito.co.nz*.

Skills you will need

- Level 1 numeracy.
- Level 1 literacy.
- The ability to work as a team.
- The ability to meet deadlines.
- A willingness to listen and learn.
- A satisfactory level of fitness and health.
- An awareness of health and safety procedures.
- A current driver's license is beneficial.

Becoming a drainlayer

Good drainlaying is an essential part of any community. The reticulation of clean drinking water and the efficient and safe removal of waste is an integral part of any infrastructure.

How do you get started?

Like the plumber and gasfitter, you need to find an employer willing to employ you as an apprentice. You learn on the job and are also required to attend block courses. The apprenticeship takes two years and at the end of that time you are awarded a National Certificate in drainlaying. Further information can be found at *http://ito. co.nz/apprenticeships/drainlaying*.

Skills you will need

- Level 1 numeracy.
- Level 1 literacy.
- The ability to work as a team.
- The ability to meet deadlines.
- A desire to work outdoors.
- A willingness to listen and learn.
- A satisfactory level of fitness and health.
- An awareness of health and safety procedures.
- A current driver's license or a heavy traffic license is beneficial.

ISBN: 9780170238670

1 Using this garage and workshop plan, calculate the length of spouting required on the two longer sides.

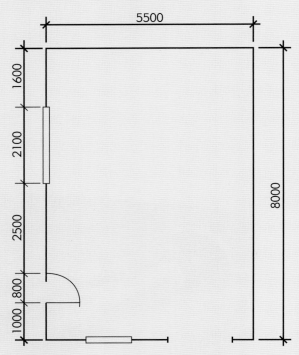

2 Using the house plan from question 6 on page 34, calculate the length of spouting required around the whole building.

3 Using the house plan and its scale from question 7 on page 34, calculate the length of spouting required around the whole building.

4 The bathroom in a house plan, with scale factor 1 : 50, is measured at 50 mm by 30 mm. What are the dimensions of the actual bathroom? (Give the answer in metres.)

5 Using the house plan on page 49, what is the area of …
   a the bathroom?
   b the kitchen?

6 Gradient is the slope of line. The formula is $\frac{rise}{run}$ or $\frac{y}{x}$
Calculate the gradient of these slopes.

a

2 m

20 m

b

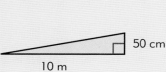

50 cm

10 m

c

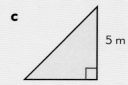

5 m

5 m

ISBN: 9780170238670

7 Using your answers in question 6 above, you can see the gradient can also be written as a ratio rise : run. Write these gradients as ratios (in simplest form).

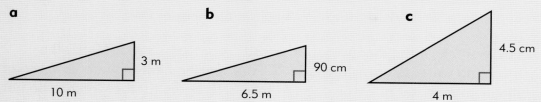

a

3 m

10 m

b

90 cm

6.5 m

c

4.5 cm

4 m

8 The ratio of the slope of a sewer pipe should be between 1 : 40 and 1 : 110, with 1 : 80 being a common value. You have to design the trench for a sewer that runs for 150 m. What should the fall be if the ratio is 1 : 80?

9 If the slope in a sewer pipe is too flat there is a risk of blockages. If the ratio is flatter than 1 : 110 this can be a danger. If a 200 m sewer line had a fall of 1.5 m do you think there might be problems? Explain why.

10 A roof in a high snowfall area of New Zealand must have a steep angle so snow will not build up on it in a snowstorm. The rise, or vertical height from ceiling to top of roof, must be 1.1 times the horizontal distance, or run. If the run is three metres, what must the rise be?

11 You are asked to design a trench that is 60 m long with a gradient of 2.5%. What must the height difference between the two ends be?

12 A long drainage ditch on a farmer's paddock is 595 m long. The paddock is 17 m higher at the top of the ditch than at the bottom. Calculate the gradient as a fraction, as a ratio, as a percentage, and as millimetres of fall per metre of run.

13 You need to wrap Teflon thread tape four times round a thread on a 50 mm diameter pipe. Calculate how much tape will be needed.

14 A toilet cistern measures 550 mm wide by 350 mm high, and its length from front to back is 240 mm. Calculate its capacity, in litres.

15 If a cylinder has a height of 1.1m and a diameter of 0.78m, calculate the …

   a total surface area of the cylinder. (Answer to 2 decimal places.)

   b the volume of the cylinder.

16 How many litres of water will fill 20 m of pipe with a diameter of 100 mm?

17 Calculate the volume of water that would be between the inner and outer cylinders of this calorifier:

   The outer radius (R) = 450 mm

   The inner radius (r) = 400 mm

   The height (h) = 900 mm

ISBN: 9780170238670

18 To calculate how long it will take to heat a hot water cylinder, use this formula:

$$T = \frac{\text{(mass of water)} \times \text{(temperature difference)} \times \text{(specific heat of water)}}{\text{Power rating of electric element in kW}}$$

For example, if the thermostat on a 180 L cylinder is set at 60 °C, the incoming water temperature is 15 °C, the specific heat of water is 4.186 and the element is set at 3 kW, it would take 3.15 hours for the water to be hot. (Note: Answer needs to be ÷ 3600 to get hours.)

a Calculate how long it would take if the heating element was only 2kW instead of 3kW.

b Another tank is a smaller, under-bench model of 25 litre capacity. The incoming water is colder now, only 10 °C and the hot water is needed at 70 °C. The element is rated at 2 kW. Calculate how long it would take to heat this tank.

Becoming an electrician

An electrician is an essential tradesperson. It is important that all electrical work be carried out by a certified electrician.

How do you get started?

Like the other trades, you need to find an employer willing to employ you as an apprentice for the duration of your apprenticeship of approximately three years. Training is done on the job and at polytechnics. The provider is the Electrician Trade in Training Organisation (ETITO). At the completion of the apprenticeship you will be awarded a National Certificate of electrical engineering Level 2. Further information can be found at *www.switchedoncareers.co.nz*.

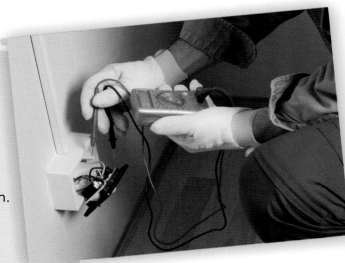

Skills you will need

- Level 1 numeracy.
- Level 1 literacy.
- The ability to work as a team.
- The ability to meet deadlines.
- A willingness to listen and learn.
- A satisfactory level of fitness and health.
- An awareness of health and safety procedures.
- A current driver's license is beneficial.

ISBN: 9780170238670

Exercise 10

1 Given the formula voltage (V) = current (I) x resistance (R):
   a What is the voltage if the current is 5A and it has a resistance of 10Ω?
   b What is the current in a circuit with a voltage of 6V and a resistance of 3Ω?
   c What is the resistance of a circuit with a voltage of 12V and a current of 2A?

2 Total resistance (Rt) is the sum of the resistors in the series:
   a What is (Rt) if R1 = 120Ω, R2 = 110Ω and R3 = 125Ω?
   b What is (Rt) if R1 = 5.25Ω, R2 = 8.95Ω and R3 = 15.9Ω?
   c What is R1 if the total resistance Rt is 27.95Ω, R2 = 12.5 Ω and R3 = 7.8Ω?

3 An apprentice needs to construct a 1.65 m² switchboard from materials of 2.4 m². How much material is left over?

4 If eight lights are connected in a series and each light uses 8.5W, how many watts in total are used?

5 The plumber has installed a new electric hot water cylinder and the electrician connects it to heat at night only. It is estimated to use 90 kilowatt-hours of electricity a week. If the night rate is 14.38 cents per kWh, what is the approximate cost each month?

6 If the cylinder in the previous question was connected to the day rate of 20.13 cents per kWh, what is the weekly saving of having it connected to the night rate?

7 A 300 watt pump runs day and night in a fish pond to make a fountain. Electricity costs 9c per unit (9c/kW hour) during the night-rate hours of 9 p.m. to 6 a.m., and 19 c/unit for the daytime-rate hours of 6 a.m. to 9 p.m. How much does it cost altogether to run the pump for a whole week of seven days and seven nights?

8 The average household in the North Island of New Zealand uses 8000 kWh per year, and in the South Island the average household uses 12.5% more. How many kWh do houses in the South Island use?

9 It is essential to know what are good insulators and conductors of electricity.
   a What are insulators?
   b What are conductors?
   c Give two examples of insulators and conductors.

ISBN: 9780170238670

10 Use the following floor plan to answer the questions below.

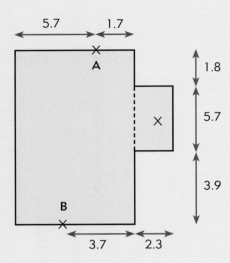

a What is the perimeter of this floor plan?

b What is the area of the floor space?

c What is the shortest distance between A and B, going around the perimeter?

d What percentage of the plan is room X?

e Continuously interconnect four power points located around the perimeter, allow an extra 0.25 m for each point. Find the total length of cable required.

f If cable costs $3.45/m and each power point costs $4.15, what is the total cost of cable and power points?

ISBN: 9780170238670

Paint, paper and plaster

Interior walls are usually clad in gibraltor (gib) board and then covered with plaster, then paint or wallpaper.

Becoming a painter or decorator

If you want to become a painter or decorator, you need to have at least three years secondary schooling and it would be useful to have studied mathematics, English, art and design, technology and workshop but not essential. The painting qualification is the National Certificate in painting (Level 3), which is overseen by the Creative Trades Industry Training Organisation (CTITO). The decorating qualification is the National Certificate of decorating.

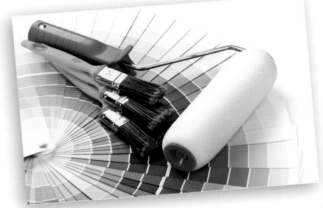

How do you get started?

Like the builder, you need to find an employer who is willing to employ you while you train. A qualified painter will oversee your practical and theory work. It will take approximately three years to complete all the required unit standards and practical work. Further information can be found at *www.decoratenz.org.nz.*

Skills you will need

- Level 1 numeracy.
- Level 1 literacy.
- The ability to work as a team.
- The ability to meet deadlines.
- A willingness to listen and learn.
- A satisfactory level of fitness and health.
- An awareness of health and safety procedures.
- A current driver's license is beneficial.

ISBN: 9780170238670

Painting

One litre of paint covers approximately 10 m². Paint comes in these sizes at the indicated average cost.

1 litre	**$50**
4 litres	**$120**
10 litres	**$185**

Exercise 11

1 How many litres of paint would be needed to cover these areas with one coat of paint? (Answer to nearest whole number.)

  a 40 m x 5 m

  b 124 m x 2.5 m

  c 86.5 m x 3m

2 What would be the cost and most economical way to buy paint for each of the areas in the previous question?

3 a What is the area of the four walls of this building?

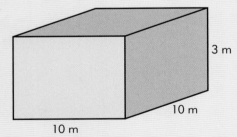

3 m

10 m

10 m

  b How many litres of paint are needed to give it two coats on the inside?

  c What is the cost and most economical way to buy the paint?

4 What is the cost and most economical way to buy paint for the walls of this building? Note that two coats are required.

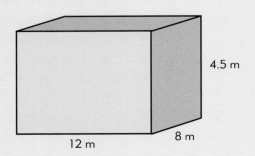

4.5 m

12 m

8 m

ISBN: 9780170238670

5 a i What is the area of the roof of this garage?

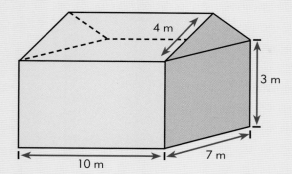

4 m

3 m

10 m

7 m

ii How much paint is required to give the roof two coats?

b i What is the area of the walls?

ii If there is 26 m² of windows and doors, how much area needs to be painted?

iii How much paint is required to give the walls two coats?

c What is the cost and most economical way to buy the paint for the whole building?

6 a If the stud height of this house is 2.4 m, what is the area of the outside walls (exclude windows and doors)?

b How much paint would be needed to give the walls three coats?

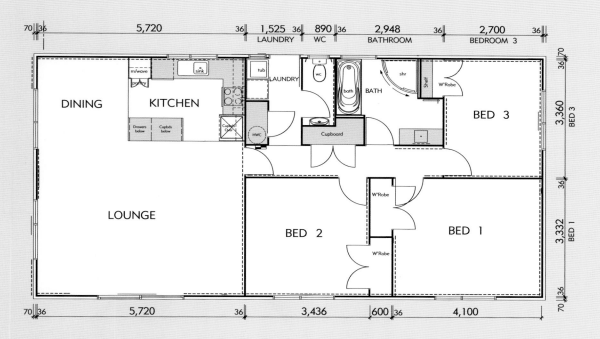

ISBN: 9780170238670

7 Peter and Sandy had a fence put around their property. The fence was 1.5 m high and needed to be painted on both sides.

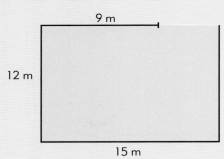

9 m

12 m

15 m

a What is the total area to be painted?

b If two coats are needed each side of the fence, how many litres of paint will be needed?

Papering

Exercise 12

1 Find the perimeter of these rooms.

a

6 m

10 m

b

3 m   1 m

8 m

7 m

c

6 m

4.8 m

4 m

1.2 m

2 m

ISBN: 9780170238670

2 Use the chart below to find out how many rolls of wallpaper are needed for the room sizes in the table that follows. The first example has been completed.

Walls		Distance around room (doors and windows included)																	
METRES		8.53	9.75	10.97	12.19	13.41	14.63	15.85	17.07	18.29	19.51	20.73	21.35	23.95	24.38	25.60	26.82	28.04	29.26
	FEET	28′	32′	36′	40′	44′	48′	52′	56′	60′	64′	68′	72′	76′	80′	84′	88′	92′	96′
2.13 to 2.29	6′10″ to 7′5″	5	5	6	6	7	7	8	8	9	9	10	10	10	11	11	13	13	14
2.30 to 2.44	7′7″ to 8 0″	5	5	6	6	7	7	8	9	9	10	10	11	11	12	13	14	14	15
2.45 to 2.59	8′1″ to 8′5″	5	6	6	7	7	8	8	9	9	10	10	12	12	13	14	15	15	15
2.60 to 2.74	8′7″ to 9′0″	5	6	6	7	7	8	9	9	10	10	11	13	13	14	14	15	15	16
2.75 to 2.90	9′1″ to 9′6″	5	6	7	7	8	8	9	10	10	11	11	13	14	14	15	15	16	17
2.91 to 3.05	9′7″ to 10′0″	6	6	7	8	8	9	10	10	11	11	12	14	14	15	16	16	17	18
3.06 to 3.20	10′1″to 11′6′	6	6	7	8	9	9	10	11	11	12	13	14	15	16	16	17	18	18

Rooms	a	b	c	d	e	f	g	h
Height of wall (m)	2.35	3.12	2.65	2.53	3.17	2.63	2.63	2.75
Distance around room (m)	15.9	17.51	14.7	9.8	11.02	25.65	29.3	18.33
Number of rolls required	8 rolls							

3 If the height of each room in 2.35 m, how many rolls are needed for the rooms in question 1?

4 What is the total wallpaper cost for the following rooms?

   a Question I a: wallpaper costs $70 per roll.

   b Question I b: wallpaper costs $72.95 per roll.

   c Question I c: wallpaper costs $69.95 per roll but was bought in sale at a 25% reduced price.

Plastering

Exercise 13

1 a Find the area of the walls of the following rooms if the height is 2.4 m.

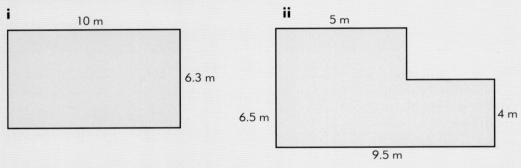

i

8 m

5.5 m

ii

5 m   2 m

6 m

9 m

b If plastering costs $40.50 per m², how much would it cost to have the walls in the previous question plastered?

2 The stud height of these rooms is 3 m.

i

10 m

6.3 m

ii

5 m

6.5 m   4 m

9.5 m

a What is the area of walls in the two rooms above?

b How much would it cost to have these rooms plastered if the plastering cost $40.50 per m²?

ISBN: 9780170238670

Exercise 14

Use this plan of an addition (stud height is 2.35 m) to answer the questions that follow.

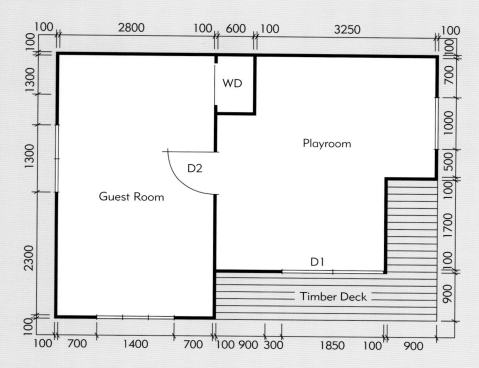

1   Do you think it is upstairs or downstairs? Why?

2   Molly and Terry decide to wallpaper the guest room. How many rolls of wallpaper do they need? (Refer to the chart on page 61.)

3   The walls of the playroom are to be plastered. How many m² are to be plastered? (Ignore doors and windows.)

4   Once the playroom is plastered it requires two coats of paint. How many litres would be required? Painting will take 20 hours.

5   They decide to stain the timber deck. How many litres of stain will they need if it requires two coats?

6   Costs for the decorations are listed below:

Wallpaper	$70.50 per roll	Painter labour	$35 per hour
Paperhanger labour	$25 per roll	Paint 1L	$50
Plasterer labour	$32.50 per m²	4L	$120
Stain	$19.95 per litre	10L	$185

(all prices GST inc)

   a   What are the total costs of decorating this addition?
   b   How much GST was included in the account?

ISBN: 9780170238670

Covering the floor

Becoming a carpet/vinyl layer or tiler

Like becoming a builder, you need to find an employer who is willing to accept you as an apprentice as well as supervise your training and learning time. Learning is done on the job. FlooringNZ is responsible for qualifications associated with the flooring industry. Further details can be found at *www.floornz.org.nz*.

Carpet

Most carpets are **3.66 m (12 ft)** wide and are cut to lengths to fit the room.

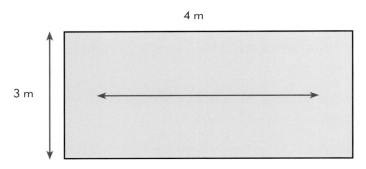

4 m

3 m

To carpet this room you would need to have two pieces of carpet, each 2 m long (so you would need a total of 4 m of carpet).

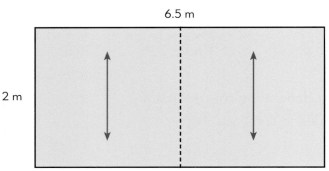

6.5 m

2 m

ISBN: 9780170238670

Exercise 15

1   How much carpet is required for the following rooms if the carpet is
    3.66 m wide? (You must follow the direction of the arrows.)

a

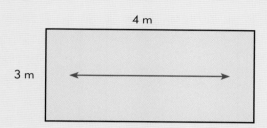

b

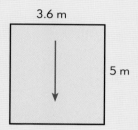

c

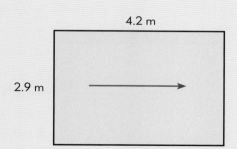

d

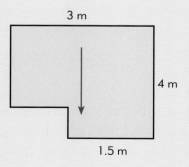

ISBN: 9780170238670

2 Copy the following diagrams into your book, select the best way to lay the carpet, then calculate how many metres of carpet would be needed.

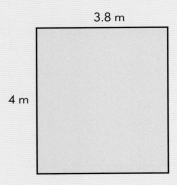

3.8 m

4 m

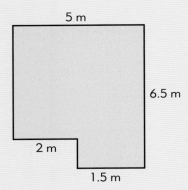

5 m

6.5 m

2 m

1.5 m

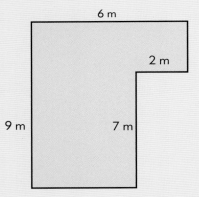

6 m

2 m

9 m   7 m

3 What would be the cost of buying carpet at $99.95 per metre for the rooms in question 1?

4 Having carpet laid includes other costs, for example rubber, foam or underfelt beneath the carpet and the cost of laying these. Using the room plans a, b and c in question 2, what would be the cost of …

i rubber underlay at $25.95 per metre?

ii carpet laying at $25 per metre?

ISBN: 9780170238670

Vinyl

Vinyl is bought by the metre – usually two or three metres wide – and is generally laid in service areas.

Exercise 16

Copy these diagrams into your book, select the best way to lay the vinyl, then calculate how many metres are required.

1

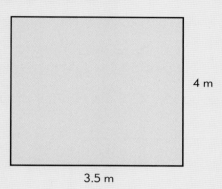

4 m

3.5 m

2

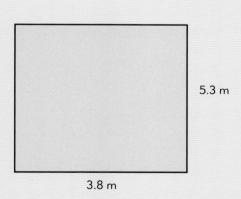

5.3 m

3.8 m

3

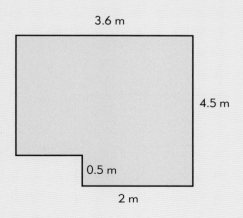

3.6 m

4.5 m

0.5 m

2 m

ISBN: 9780170238670

Tiles

Tiles come in a variety of sizes from 30 mm by 30 mm to 300 mm by 300 mm, and can be laid on the floor, walls and bench tops.

Exercise 17

1 What is the area of these tiles? (Answer in m^2.)

   a 300 mm by 300 mm

   b 200 mm by 200 mm

   c 100 mm by 100 mm

2 How many tiles of the sizes in question 1 would be required to cover 1 m^2?

3 Calculate the area of these spaces and the number of tiles needed.

   a

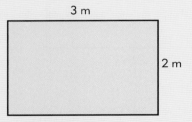

3 m

2 m

with 300 mm x 300 mm tiles

   b

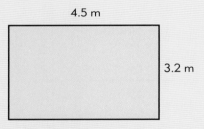

4.5 m

3.2 m

with 300 mm x 300 mm tiles

   c

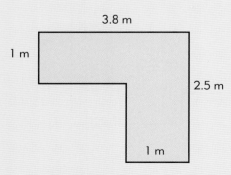

3.8 m

1 m

2.5 m

1 m

with 100 mm x 100 mm tiles

ISBN: 9780170238670

4 Use the plan below to answer the questions that follow.

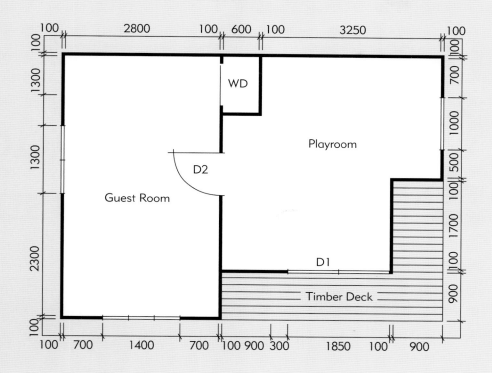

a Calculate how much carpet would be needed for the guest room (carpet is 3.66 m wide).

b Calculate how much vinyl would be needed for the playroom (vinyl is 2 m wide).

c Costs (inc. GST): Carpet $135.50 per m; vinyl $75 per m; rubber underlay $28.50 per m; labour for carpet/vinyl $25 per hour for ten hours.

  i What is the total cost of the flooring?

  ii How much is the GST?

ISBN: 9780170238670

Window dressing

Track length is usually longer than the width of the window by 20 cm each side. Curtain drop is usually from the top of the track to 20 cm below the window sill or to the floor. (Don't forget to allow for hems.)

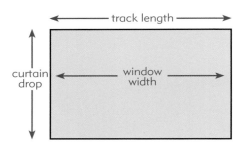

So, for example, a window measuring 200 cm x 200 cm would have a track width of 240 cm and a 220 cm curtain drop.

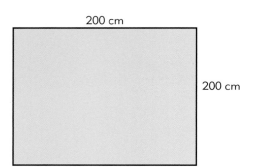

ISBN: 9780170238670

Exercise 18

1 Compare the price of readymade curtains with that of made-to-measure curtains.

2 a Find the track length and curtain drop for each of the following windows.

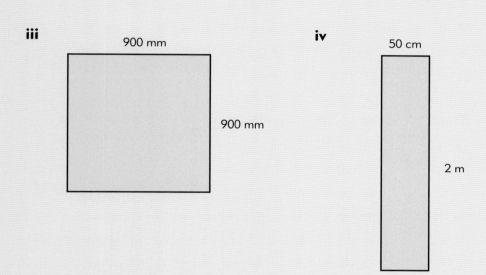

i

2 m

1 m

ii

2.5 m

1.5 m

iii

900 mm

900 mm

iv

50 cm

2 m

b To calculate the width of material required for curtains, we usually use twice the width of the window. What is the width of material required to dress each of the windows in question 2a?

c Material is usually 120 cm wide, so rounded up to the nearest whole number how many widths (drops) of material would be required in question 2a?

d Using the drops in question 2c and the drop length, calculate how much material would be needed for each window in question 2a.

e Using your answers in question d, and the cost of material at $49.95 per metre, how much would the material for each window cost?

ISBN: 9780170238670

3 **a** Estimate the dimensions of the windows in this plan.

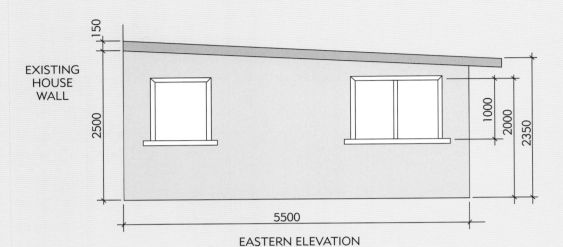

EXISTING
HOUSE
WALL

150
2500
5500
1000
2000
2350

EASTERN ELEVATION

b Using the procedure in question 2, calculate …

i what lengths of track will be required.

ii how many drops will be required.

iii how many metres of material will be required.

iv If the material costs $52.95 per metre, what is the total cost of the material?

4 Using the first three window sizes in question 2a, what price (with the discount) would you pay for readymade curtains according to the following advertisement?

ISBN: 9780170238670

10

Swimming pools and spas

- What is the difference between a swimming pool and a spa pool?
- What are the extra costs involved in maintaining either a swimming pool or a spa pool?
- Are there any special regulations required for the installation of a swimming pool or for a spa pool?

Exercise 19

1 Assuming each is 1 m deep, what is the volume of these swimming pools?

a

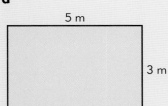

5 m

3 m

b

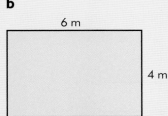

6 m

4 m

c

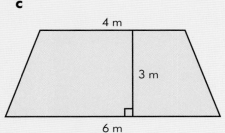

4 m

3 m

6 m

2 How many litres of water will the pools in question 1 hold if full?
(1 m³ =1000 litres)

ISBN: 9780170238670

3 Calculate the volume and capacity of the following pools.

a

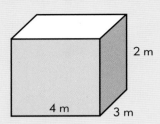

b

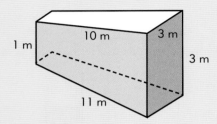

c

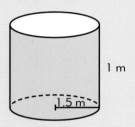

d

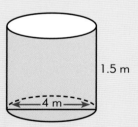

e

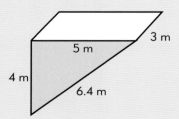

ISBN: 9780170238670

f

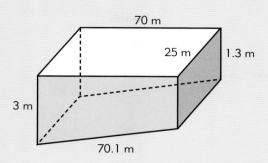

70 m
25 m
1.3 m
3 m
70.1 m

4 Calculate the volume and capacity of a pool near your school.

5 a The inside of the pools in question 3 need painting. What is the area to be painted?

   b If a litre of paint covers 10 m², how many litres are needed to give each pool interior two coats? (Answer to nearest litre.)

6 The shaded area around the pool (below) is to be paved.

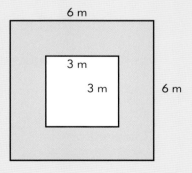

6 m
3 m
3 m
6 m

   a What is the area of the shaded part?

   b If the tiles are 300 mm x 300 mm, how many tiles would be needed for the shaded area?

ISBN: 9780170238670

7 The shaded area around the pool (below) is to be paved.

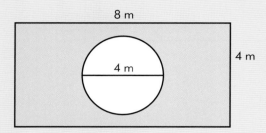

a What is the area of the shaded part? (Answer to 1 decimal place.)

b If the tiles are 300 mm x 300 mm, how many tiles would be needed for the shaded part?

8 The shaded area is the deck of this complex.

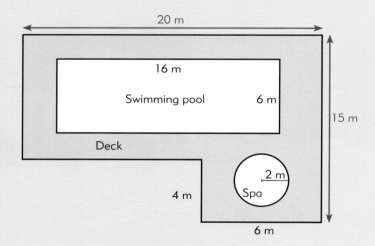

a Calculate the deck area.

b If decking costs an average of $25 per m², how much would the deck cost?

c Calculate the length of safety fencing required for this complex.

9 If a tap flows at 50 litres per minute, how long will it take to fill these pools?

a The spa pool in question 8 (depth 1.5 m).

b A pool 25 m by 10 m by 3 m pool.

ISBN: 9780170238670

Structures

- What is a structure?
- What does the dictionary say?
- How many can you see around you?

Bridges

- Why do we have bridges?
- How many types of bridges can you think of?

Exercise 20

1 a Using cardboard strips, two books and some coins, investigate the strength of your 'cardboard bridge' by seeing how many coins it will support in the middle, compared with the number of coins used on the ends.

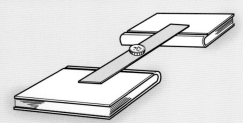 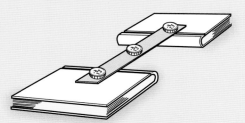

b Copy and complete this table.

Number of coins on each end	0	1	2	3	4	5	6	7	8	9	10
Maximum number of coins supported in the middle											

ISBN: 9780170238670

c Write 2–4 sentences about your findings.

d Research different types of bridges and decide what sort they are: suspension, arch, beam, cantilever, swing or viaduct. Are there any famous bridges near you?

e With four cardboard strips, some coins and books or similar objects for supports, design a bridge to support the most coins. (You may not use glue, sticky tape or pins.) Write a few sentences about your structure.

Kites

- When and where were kites invented?
- How can kites be made?

Exercise 21

Making a tetrahedral or pyramid kite

1 You will need six plastic straws, three lengths of string (one of these about five straws in length, and the other two at least two straws in length), and tissue paper.

    a Thread the long string through three straws and tie the string so that the straws make an equilateral triangle.

    b Tie one end of each of the other strings to one of the other two corners so that each corner has a long string attached.

    c Thread these strings through the remaining straws and tie at the top to make a tetrahedron.

    d Cut tissue paper to cover two faces, making sure there is sufficient overlap for glueing.

    e Place your pyramid together with other pyramids to form a large pyramid.

    f Decide where the ties are to be placed and fly your kite.

    g Investigate other kite designs.

    h A windsock is a type of kite. Construct one and position it near the classroom.

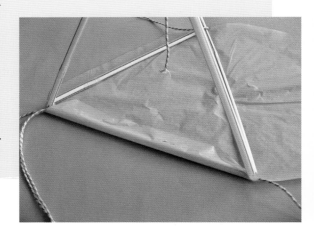

ISBN: 9780170238670

Flying objects

Exercise 22

1 **a** Using a piece of A4 paper, design your own paper dart.

   b In a clear enclosed space such as the hall or gymnasium, throw three darts and measure the distance flown.

   c Graph either best throw, or average distance, onto a histogram.

   d What is the average range of the distances?

   e Write a few sentences about your findings.

Tables and chairs

- Tables and chairs have been around a long time. What do you think the first table and chairs looked like?
- How many different types of tables and chairs are in your household and classroom?

Exercise 23

1 **a** Write a few sentences about the table and chair you are sitting at. Is it comfortable?

   b Draw a scale drawing of your ideal desktop and write an explanation for your design.

   c If you owned a restaurant, what table shapes would you choose and why?

   d Simon had a restaurant. He used only square tables and his seating arrangements were like this:

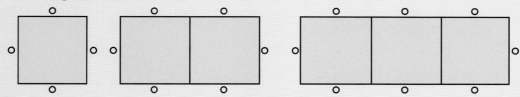

     i Draw the next two table arrangements.

     ii Copy and complete the following table.

Tables (T)	1	2	3	4	5	6	14	n
Chairs (C)								

     iii Write a rule: C =

     iv How many chairs would you need for 11 tables?

     v If Simon had 96 guests in his restaurant, how many tables and chairs would he need? Would a single row be the best use of the space? Can you design a better arrangement?

ISBN: 9780170238670

Maths in the Real World

Towers

- What is the highest building in your area?
- What is the highest building in New Zealand?
- What is the highest building in the world?
- What special features do these very high buildings have?
- Are there any special construction needs for these buildings?
- What mathematical skill would you use to measure the height of a building?

Exercise 24

1 a Find the heights of the ten tallest buildings in the world and draw a pictograph to show your findings.

b If the height of one storey of a building is approximately 3 m, how high are the following buildings?

   i   A ten storey building.

   ii A 47 storey building.

c If each storey is 3.5 m high, how many storeys are there in the following buildings?

   i   A 52.5 m building.

   ii A 126 m building.

d There are always storeys underground. If a 12 storey building has two of those levels underground, what is the ratio of storeys above ground to those underground? (Simplify your answer.)

e Using the trigonometry ratios on page 11, find the height of the buildings below. (Answer to 1 decimal place.)

   i

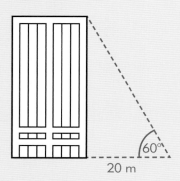

60°

20 m

ISBN: 9780170238670

ii

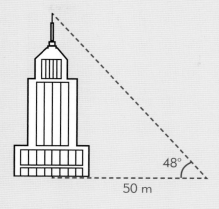

48°

50 m

iii

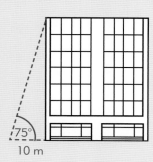

75°

10 m

iv

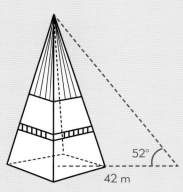

52°

42 m

f i Town planners have to be aware of the shadows buildings create, and the possible effect these shadows can have. What are some of these effects?

ii Calculate the length of the shadow the following buildings will make when the sun is at the angle stated. (Answer to 1 decimal place.)

a

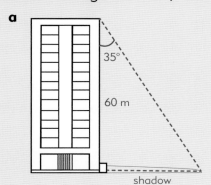

35°

60 m

shadow

ISBN: 9780170238670

b

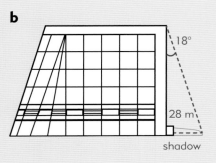

18°

28 m

shadow

c

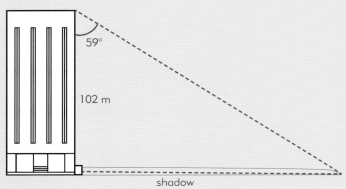

59°

102 m

shadow

g Choose a building, tree, flagpole or similarly tall object around your school and calculate its height using a tape measure and clinometer. Write a few sentences about your procedure and findings.

Letterboxes

- Who needs a letterbox?
- Do all countries use letterboxes?
- What happened with mail in New Zealand 50 years ago?

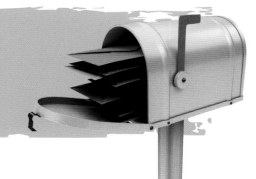

Exercise 25

1 **a** What are the characteristics of a good letterbox?

b What are the differences between a rural and an urban letterbox?

c Using cardboard, sticky tape, glue and pins, design a letterbox (rural or urban) that has all the necessary characteristics.

ISBN: 9780170238670

Stairs and steps

- Are there any stairs in your school or house?
- What sort are they?
- Do you have any buildings in your area with characteristic stairways?

Exercise 26

1 **a** Using trigonometry ratios, calculate the size of the two angles marked x.

i

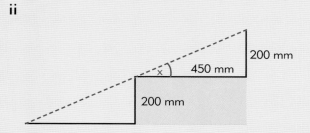

ii

b Using the trigonometry ratios, calculate the two lengths marked x (to 3 significant figures).

i

ii

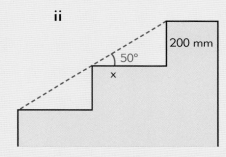

c Use this cross-section of a stairway to answer the questions on the next page.

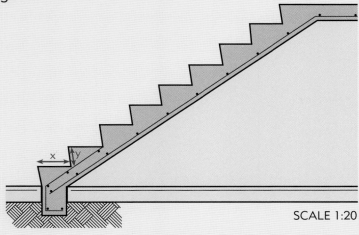

SCALE 1:20

ISBN: 9780170238670

    i What is the width (x) of the step?

    ii What is the height (y) of the step?

    iii What is the vertical height of the stairway?

    iv What is the slope (angle) of the stairway?

d Using the measurements 200 mm deep and 250 mm high, how many steps would you need in a two storey house if the vertical height is 2.5 metres?

Gates

- Gates can be made from a variety of materials and have a wide range of uses.
- What are some of these materials and uses?

Exercise 27

1 **a** Calculate how much timber you would need to make this gate. Note that X is the midpoint of AB. (Answer to the nearest metre.)

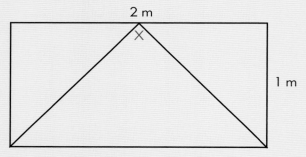

2 m

1 m

b Calculate how much timber you would need to make this gate. What is the size of angle x?

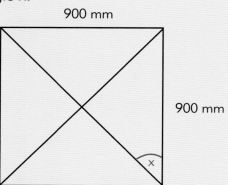

900 mm

900 mm

x

c Refer to the following table to answer the questions below.

Standard size gates			
Design code	Opening width	Height	$ (exc GST)
FPG/A01	900	1200	294.00
FPG/A02	900	1400	336.00
FPG/A03	900	1200	324.00
FPG/A04	900	1400	354.00
FPG/A05	900	1400	336.00

FPG/A01

FPG/A02   FPG/A04

FPG/A03   FPG/A05

   i What is the unit of measurement?

  ii What is the size of FPG/A04?

 iii What is the price of FPG/A02 including GST?

  iv Why do you think FPG/A04 is the most expensive gate?

   v What is the area of FPG/A03 in m²?

d i What is the maximum height of these gates?

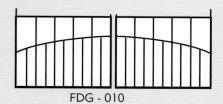

FDG - 010

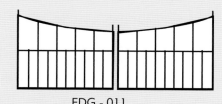

FDG - 011

Refer to the following table to answer the questions below.

Cat D	Width/ Height	Up to 2500	Up to 3000	Up to 3500	Up to 4000	Up to 5000
		Price				
Entire gates box	1200	1255.00	1315.00	1385.00	1488.00	1600.00
section frame. No	1500	1305.00	1360.00	1425.00	1520.00	1640.00
round tubes used at all	1800	1365.00	1420.00	1490.00	1585.00	1705.00

  ii What is the cost of 1500 x 3500 gates?

 iii What is the area in m² of 1200 x 5000 gates?

  iv If these prices are for a pair of gates, what is the cost of one 1800 x 4000 gate?

ISBN: 9780170238670

Putting it into practice

Achievement Standard 91026: Apply measurement in solving problems

Internal Assessment • 4 credits

NA 6:1   Apply direct and inverse relationships with linear proportions.
NA 6:2   Extend powers to include integers and fractions.
NA 6:3   Apply everyday compounding rates.

Landscaping the backyard

Molly and James decide to landscape their backyard. The features they want to install are:

- BBQ area
- Vegetable garden
- Petanque court
- Paths
- Lights

These have been shown on the scale drawing below.

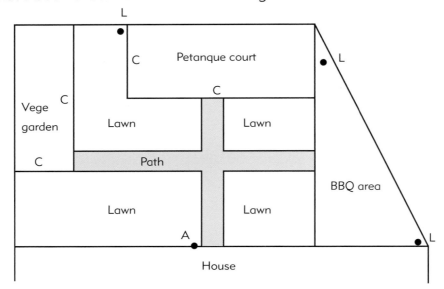

Scale   1 : 200

ISBN: 9780170238670

The fence around the three sides of the backyard is satisfactory but it does need to be painted with two coats on the inside only. It is 1.5 m high.

Concrete block edging needs to be installed on the sides marked C of the Petanque court and the garden.

The BBQ area needs to be paved with paving stones, and the pathways, which are 1 m wide and 10 cm deep, need to be concreted.

Three lights need to be installed by a certified electrician at the points marked L, with connection for each light coming from A.

Lawn area needs to be sown in grass seed.

Petanque area needs to be covered with special surface.

Costs (inc GST)

Fence paint	$125 for 10 L, $90 for 4 L and covers approximately 100 m^2
Concrete blocks	$2.50 each and are 500 mm in length
Paving stones	$3.25 each and are 300 mm x 300 mm
Concrete	$215/m^3
Lights	$39.95 each
Lighting cable	$13.50/m
Lawn seed	$8/m^2
Petanque surface	$4.30/m^2

Labour
Electrician is 8 hours @ $50/hr
Builder is 45 hours @ $38.50/hr

- **Builder A** gives them a quote for all the materials and labour but he is able to get a 12% trade discount off the materials.

- **Builder B** gives them a quote for all the materials and labour but he can get a 25% discount off only the blocks, concrete and paving stones.

Note: both builder's prices include the painter's labour.

Your task

Decide which is the better option, Builder A or Builder B, and explain why. Your work needs to be clearly set out, showing all your workings.

ISBN: 9780170238670

Achievement Standard 91030: Solve measurement problems

Internal Assessment • 3 credits

GM 6:2   Apply the relationships between units in the metric system, including the units for measuring different attributes and derived measures.

GM 6:3   Calculate volumes including prisms, pyramids, cones and spheres, using formulae.

Pool renovations

Claudia and David need to renovate their pool complex. The diagrams below show the size and shape of the pool from above and side on. The inside of the pool needs to be water blasted, sanded and repainted with two coats of paint. They also need to renew the 1 m concrete path around the edge of the pool. This path is 10 cm deep.

It is estimated the renovations will take 60 hours, including one day to do the water blasting. How much will it cost to fill the pool if the council charges $13 per KL of water?

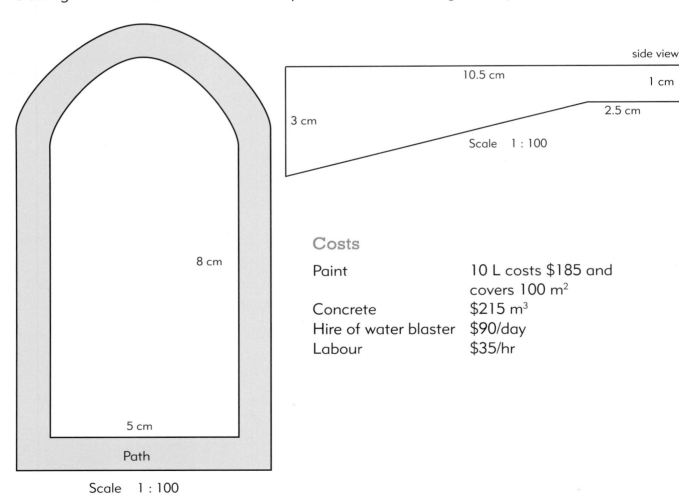

side view

10.5 cm   1 cm

3 cm   2.5 cm

Scale   1 : 100

8 cm

5 cm

Path

Scale   1 : 100

Costs

Paint	10 L costs $185 and covers 100 m²
Concrete	$215 m³
Hire of water blaster	$90/day
Labour	$35/hr

Your task

Calculate the total costs of the renovations and how much it will cost to fill the pool, ready for the summer. Your work needs to be clearly set out, showing all your workings.

Achievement Standard 91032: Solve measurement problems involving right-angled triangles

Internal Assessment • 3 credits

GM 6:1   Measure at a level of precision appropriate to the task.

GM 6:5   Recognise when shapes are similar and use proportional reasoning to find an unknown length.

GM 6:6   Use trigonometric ratios and Pythagoras' theorem in two and three dimensions.

Staircase installation

A set of stairs need to be installed in a house connecting the ground floor with the first, which has a stud height of 2.5 m.

A standard step is 200 mm deep and 250 mm high.

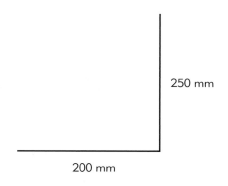

250 mm

200 mm

Your task

Use Pythagoras and trigonometry to design a safe staircase, showing the angle of elevation and other measurements associated with its installation. You will need to draw a scale drawing of your staircase and write a summary of your design.

ISBN: 9780170238670

Enrichment

1   List the occupations involved in the construction industry. How dependent is your local community and economy on the construction industry?

2   Arrange some work experience in one of the occupations involved in construction.

3   Invite a person involved in the construction industry to speak to you and others.

4   Investigate the various forms of insulation and what value these provide in a building.

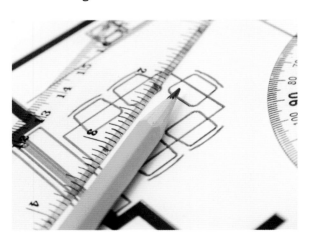

5   Investigate the pros and cons of various forms of heating available for houses.

6   Electricity costs are a significant cost in any household. Investigate the various options available from several electricity providers.

7   A representative of your local power authority could visit and describe the savings that can be made through power efficiency.

8   You may want to investigate some of the other trades involved in the construction industry such as surveying, architecture, draughting, cabinet making or landscape gardening.

9   Draw a flow chart to show the stages involved in building a house. Here are some examples which may help: Painting the interior; Putting on the roof; Carpet laying; Surveying the site; Getting a permit; Buying a section; Digging the footings.

ISBN: 9780170238670

10 Design and draw to scale your dream home.

11 Car parking areas are essential and need to be carefully planned. Design a staff/student car parking area for your school.

12 Investigate the advantages and disadvantages of renting a three bedroom house instead of buying one over a five-year period.

13 You are a real estate salesperson and have been asked to advertise a house. Prepare and design the advertisement.

14 Metric paper sizes are based on a rectangular piece of paper of area 1 m² called A0. All other sizes are half sheets of the previous size.

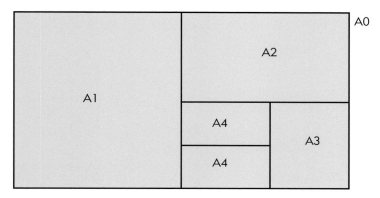

a Copy and complete this table (round sensibly).

Paper size	Dimensions (mm)	Area (mm²)
A0	841 x 1189	= 1 000 000
A1	594 x 841	
A2	420 x 594	
A3	297	
A4	210	

b How many A4 sheets can be cut from one piece of A0 paper?

15 Investigate the 'Golden Rectangle' and where in life it is applied.

16 Construct a copy of this rectangle and its shapes. Cut out the pieces and rearrange them to form a square.

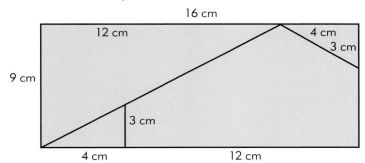

17 Draw a 10 cm by 6 cm rectangle. Now inscribe a triangle inside the rectangle so that its area is 50% of the rectangle. Write a report on your findings.

ISBN: 9780170238670

Answers

Exercise 1

1
a 0.5, 50%
b 0.6, 60%
c 0.875, 87.5%
d 0.333, 33.3%
e 435.5%

2
a 1¾, 175%
b 1/20, 5%
c ⅖, 40%
d 2 11/20, 255%
e 9 6/25, 924%

3
a 7/20, 0.35
b 2½, 2.5
c $\frac{1}{8}$, 0.125
d 3/20, 0.15
e 1/20, 0.05

4
a 256 mm²
b 30.25 m²
c 0.1225 m²
d 6.25 m²

5
a 25 mm
b 12.5 m
c 6.928 m
d 34.641 mm

6
a 1000 mm
b 5250 mm
c 2.3 m
d 560 mm
e 1.55 m
f 0.04 m
g 6800 mm
h 4050 mm
i 1.65 m
j 560 mm

7
a 1 kg
b 2500 g
c 1000 kg
d 40 000 g

8
a 1000 mL
b 250 mL
c 2.5 L
d 0.015 KL

9
a P = 16 m A =16 m²
b P = 8000 mm A = 3 750 000 mm²
c P= 30 m A = 71.2 m²
d P = 60m A = 150 m²

10
a 27 000 000 mm³
b 87 500 000 mm³
c 84.375 m³
d 2000 cm³

11
a 2826 mm, 953 775 000 mm³, 0.953775 m³
b 314 mm, 1 570 000 mm³, 0.00157 m³
c 1758.4 mm, 233 867 200 mm³, 0.2338672 m³

12
a 953.8 L
b 1.57 L
c 233.87 L

13
a 2 kg cement, 1 kg water, 10 kg builders mix
b 8 kg cement, 4 kg water, 40 kg builders mix
c 260 kg

14
a 11 kg
b 7.5 gal
c 3.2 km
d 30.48 cm

15 25.4 cm

16 9 kg

17
a 32.2 °C
b 100 °C
c -17.8 °C
d 73.9 °C

Exercise 2

1 43

2 Lot 46, Right of way, driveway to section

3 Lot 39 and 72

4 600 m²

5 Lot 39 corner site and larger

6

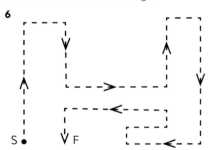

7 2.9655 ha

8 690m² .0690 ha

9 2 : 5

Exercise 3

1 $ 404 000

2
a 38 4%
b 36 0%
c 45 -2%

3
a $455 000
ii $429 000
b Larger corner site

4
a $2325.40
b $2730.77
c $2612.90

5
a $229.65
b Lot 152
c i $3658.54
ii $3215.82
d $18 500
e i $596 800
ii 71%

f Resolution 4, Endeavour 4, Discovery; wanting a certain standard of houses

6
a $11 060
b $17 250
c $23 175
d $6122.50
e $20 720

7
a Queenstown
b location, tourism
c Dunedin
d location, demand
e Wellington 23.5%

ISBN: 9780170238670

f

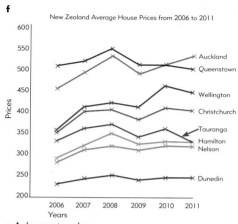

New Zealand Average House Prices from 2006 to 2011

g Ask your teacher

8 **a** Auckland Central–Eastern Suburbs
b Auckland South–Papatoetoe
c Prior to economic downturn and supply and demand
d 12.8%
e

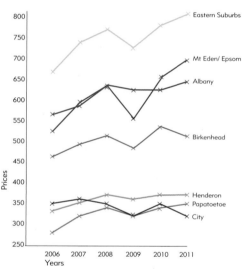

Median House Prices for Auckland from 2006 to 2011

9 **a** Christchurch Seaside
b 14.2%
c $417 400
d Christchurch South West Inner 24.4%
e Unaffected by the earthquakes

Exercise 4

1 **a** 433.2 m² **b** 592 m² **c** 326.6 m² (1 dp)
d 441.6 m² (1 dp) **e** 749.16 m² (2 dp)

2 **a** 625.5 m² (2 dp) **b** 208.4 m² (1 dp) **c** 139 m²

3 **a** 100 m **b** 89.2 m **c** 107 m

4 **a** 2100 m² **b** 7410 m² **c** 10 435 m²
d 10 205 m²

5 1

6 **a**

100 × 100	10 000	400 m
250 × 40	10 000	580 m
200 × 50	10 000	500 m
400 × 25	10 000	850 m
1000 × 10	10 000	2200 m

b **i** 1000 × 10 **ii** 100 × 100
iii square = smallest perimeter

7 **a** 326.8 m **b** 222.9 m

8 Ask your teacher

9 **a** 1000 m² **b** 144 m² **c** 14 4%
d 856 m²

10 **a** 120 m² **b** 720 m² **c** $\frac{1}{7}$

11 **a** 3234.2 m² **b** 8962.2 m²

12 **a**

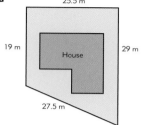

25.5 m
19 m
House
29 m
27.5 m

b 612 m² **c** 6 m

Exercise 5

1 **a** toilet **b** door **c** wardrobe
d stove **e** window **f** stairs down

2 **a** 500 mm **b** 2500 mm **c** 5000 mm
d 12 500 mm **e** 1250 cm **f** 600 cm

3 **a** 1200 mm **b** 4500 mm **c** 1800 mm
d 6900 mm **e** 4050 mm **f** 8100 mm

4 5500 on the plan = 5.5 cm, 8000 on the plan = 8 cm

5 **a** 2.40 m **b** 1.60 m **c** 3.84 m²
d 1.60 m²

6 **a** a = 5.1 m, b = 6.8 m, c = 5.5 m, d = 2.9 m,
e = 2.9 m, f = 1.2 m **b** window **c** 29 m²

7 **a** 75.9 m² **b** 15.6 m² (1 dp) **c** 19.3%

8

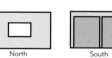 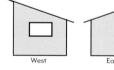
North South West East

9 **a** 96.2 m² **b** 14.5% **c i** 9.76 m × 7.4 m
ii 4.3 m × 3.24 m **iii** 3.1 m × 3.24 m

Exercise 6

1 **a** 60 m³ (0.06 m³) **b** 0.4 m³ **c** 2.97 m³
d 3.875 m³ **e** 0.99 m³
f 1.52 m³

2 **a** $14 100 **b** $94 **c** $697.95
d $910.63 **e** $232.65 **f** $357.20

3 **a** 2 m³ **b** $470

4 **a** 10 **b** 63 **c** 215
d 184 **e** 257

5 **a** 15 m³ $3525 **b** 22.2 m³ $5217
c 23.94 m³ $5625.90 **d** 25.42 m³ $5973.70
e 25.35 m³ $5957.25

6 **a** 41 **b** 73

7 **a i** 1000, 2200, 4000, 5000, 6500 **ii** 7800
b i 1200, 1300, 2500, 2500, 1700, 300 **ii** 9500
c i 1600, 1100, 1000, 1500 **ii** 1500, 5000
iii 7500

ISBN: 9780170238670

8 **a** 6 **b** 4 **c** 9

9 **a** 576 mm **b** 543 mm **c** 592.5 mm

Exercise 7

1 **a** 4.1 m **b** 5.2 m **c** 3.9 m
d 3.6 m

2 **a** 11.2 m **b** 15.9 m

3 **a** 22.6 m **b** 20.6 m **c** 27.8 m
d 26.8 m

4 **a** 9.0 m **b** 4.5 m **c** 9.6 m
d 13.2 m

5 **a** 36.9° **b** 55.0° **c** 7.6°
d 2.3°

6 **a** 125 m **b** 297 m **c** 528 m
d 688 m

7 **a** 3.3 m **b** 1.15 m **c** 6 m
d 25° **e** 218 m **f** $7383.75

8 **a** 138 600 mm², 0.1386 m² **b i** 455
ii 740 **iii** 971 **iv** 758
v 1339

Exercise 8

1 **a** 2430 **b** 778 **c** 3159
d 171 **e** 6075

2 **a** 0.01748 m² **b** 0.1748 m² **c** 6.118 m²
d 21.6752 m² **e** 64.3264 m² **f** 51.968 m²

3 17 m² or 16 m²

4 **a** 3 or 4 **b** 11 or 12 **c** 15 or 16

5 **a** Excluding overlap **b i** 205 mm **ii** 175 mm
iii 14.6%
c i 175 mm **ii** 12 **iii** 50.4 m
iv $878.22

6 **a** 180 m **b** $6974.10

7 155 m

8 **a** 10
b i 2300 mm, 1700 mm **ii** 1500 mm **iii** 1100 mm
c 21 m²

Exercise 9

1 **a** 16 000 mm

2 30.2 m

3 36.2 m

4 2.5 m x 1.5 m

5 **a** 6 m² **b** 7.2 m²

6 **a** 0.1 **b** 0.05 **c** 1

7 **a** 3 : 10 **b** 9 : 65 **c** 9 : 8

8 1.875 m

9 There could be problems because there should be at least
1.8 m for this length

10 3.3 m

11 1.5 m

12 1/35, 1 : 35, 2.86%, 28.6 mm/m

13 628 mm

14 46.2 L

15 **a** 3.65 m² **b** 0.526 m³

16 157 L

17 120.2 L

18 **a** 4 hr 42 mins **b** 52 mins

Exercise 10

1 **a** 50V **b** 2A **c** 6Ω

2 **a** 355Ω **b** 30.1Ω **c** 7.65Ω

3 0.75 m²

4 68W

5 $51.77

6 $5.18

7 $25.62

9 9000 kWh

10 **a** Insulators are materials that do not respond to electricity
and resist the flow of electrical charge
b Conductors allow electrical charges to pass through them
c Insulators: Glass, wood, rubber, plastic, air. Conductors:
Copper, silver, gold

11 **a** 42.2 m **b** 97.47 m²
c From A to B going anticlockwise 20.8 m **d** 13.5%
e 43.2 m **f** $165.64

Exercise 11

1 **a** 20 L **b** 31 L **c** 26 L

2 **a** 2 x 10 L, $370 **b** 3 x 10 L, 1 x 1 L, $605 **c** 3 x 10 L,
$555.00

3 **a** 120 m² **b** 24 L **c** 2 x 10 L, 1 x 4 L, $490

4 4 x10 L, $740

5 **a i** 80 m² **ii** 16 L **b i** 102 m²
ii 76 m² **iii** 15.2 L
c 3 x 10 L, 2 x 1 L, $655

6 **a** 99.84 m² **b** 30 L

7 **a** 144 m² **b** 29 L

Exercise 12

1 **a** 32 m **b** 30 m **c** 24 m

2 **b** 11 **c** 9 **d** 7
e 7 **f** 15 **g** 16
h 11

3 **a** 16 **b** 15 **c** 12

4 **a** $1120 **b** $1094.25 **c** $626.85

Exercise 13

1 **a i** 64.8 m² **ii** 81.6 m²
b i $2624.40 **ii** $3304.80

2 **a i** 97.8 m² **ii** 96 m²
b i $3960.90 **ii** $3888

Exercise 14

1 Upstairs (stairs down)

2 8

3 37.6 m²

4 8 L

5 2 L

6 **a** Costs of addition: Wallpaper 8 x $70.50 = $564;

ISBN: 9780170238670

Plastering 37.6 m x $32.50 = $1222; Paint 10 L $185;
Stain 2 L $39.90; Painter $700; Paperhanger $200;
Total cost = $3165.90

b $412.94

Exercise 15

1 **a** 4 m   **b** 5 m   **c** 4.2 m   **d** 4 m

2

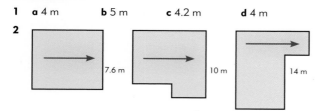

7.6 m   10 m   14 m

3 **a** $399.80   **b** $499.75   **c** $419.79
   d $399.80

4 **a i** $197.22   **ii** $190   **b i** $259.50
   ii $250   **c i** $363.30   **ii** $350

Exercise 16

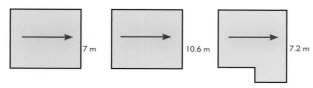

7 m   10.6 m   7.2 m

Exercise 17

1 **a** 0.09 m²   **b** 0.04 m²   **c** 0.01 m²

2 **a** 12   **b** 25   **c** 100

3 **a** A = 6 m², 67   **b** 14.4 m², 160   **c** 5.3 m², 530

4 **a** 4.9 m   **b** 12.15 m
   c i Carpet $663.95, vinyl $911.25, rubber underlay
$139.65, labour $250: Total cost $1964.85   **ii** $256.28

Exercise 18

2 **a i** 240 cm, 120 cm   **ii** 290 cm, 170 cm
   iii 130 cm, 110 cm   **iv** 90 cm, 220 cm
   b i 4 m   **ii** 5 m   **iii** 1.8 m   **iv** 1 m
   c i 4   **ii** 5   **iii** 2   **iv** 1
   d i 4.8 m   **ii** 8.5 m   **iii** 2.2 m   **iv** 2.2 m
   e i $239.76   **ii** $424.58   **iii** $109.89
   iv $109.89

3 **a** 1000 mm x 1000 mm, 1800 mm x 1000 mm
   b i 140 cm, 220 cm   **ii** 2, 3   **iii** 6 m   **iv** $317.70

4 **i** $114.95 $91.96   **ii** $149.95 - $119.96
   iii $74.95 -> $59.96

Exercise 19

1 **a** 15 m³   **b** 24 m³   **c** 15 m³

2 **a** 15 000 L   **b** 24 000 L   **c** 15 000 L

3 **a** 24 m³, 24 000 L   **b** 60 m³, 60 000 L   **c** 7.065 m³, 7065 L
   d 18.8 m³, 18 800 L   **e** 30 m³, 30 000 L
   f 3762.5 m³, 3 762 500 L

5 **a i** 40 m²   **ii** 85 m²   **iii** 16.5 m²
   iv 31.4 m²   **v** 51.5 m²   **vi** 2161 m²
   b i 8 L   **ii** 17 L
   iii 3.3 L   **iv** 6.3 L   **v** 10.3 L
   vi 432.2 L

6 **a** 27 m²   **b** 300

7 **a** 19.4 m²   **b** 216

8 **a** 135.4 m²   **b** $3385   **c** 70 m

9 **a** Approx 6 hr 20 min   **b** 250 hr

Exercise 23

1 **d i**

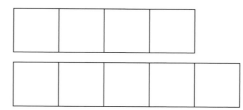

ii

Tables (T)	1	2	3	4	5	6	14	n
Chairs (C)	4	6	8	10	12	14	30	2n +2

iii C= 2t +2   **iv** 24   **v** 47 tables, 96 chairs

Exercise 24

1 **b i** 30 m   **ii** 141 m
   c i 15   **ii** 36
   d 5 : 1
   e i 34.6 m   **ii** 55.5 m   **iii** 37.3 m
   iv 53.8 m
   f i Ask your teacher **ii** A = 42 m, B = 9.1 m, C = 169.8 m

Exercise 26

1 **a** 33.7°   **ii** 24.0°
   b i 336 mm   **ii** 168 mm
   c i 280 mm   **ii** 180 mm   **iii** 1580 mm
   iv 35°
   d 10

Exercise 27

1 **a** 8.83 m   **b** 6.15 (8.83 / 6.15) m, 45°
   c i millimetres (mm) **ii** 900 mm x 1400 mm
   iii $386.40   **iv** Ornate   **v** 1.08 m²
   d i 1800 mm   **ii** $1425   **iii** 6 m²
   iv $792.50

Putting it into practice: Achievement Standard 91026

Costs

Fence area: 50.1 x 2 = 100.2 m² \ 1 x 10 L and 1 x 4 L = $215

Concrete blocks: 31 m \ 62 blocks @ $2.50 each = $153

Paving stones: 400 @ $3.25 = $1300

Concrete: 1.88 m³ @ $215/m² = $404.20

Lights: 3 @ $39.95 = $119.85

Lighting cable: 18.2 m @ $13.50/m = $245.70

Lawn seed: 103 m² @ $8/m² = $824

Petanque area: 40 m² @ $4.30/m² = $172

Labour

Electrician $400

Builder $1732.50

Total price: $5566.25

ISBN: 9780170238670

Builder A

Materials:	$3433.75 less 12% = $3021.70
Labour:	$2132.50
Total quote for A:	$5154.20

Builder B

Blocks, paving stones, concrete:	$1857.20 less 25% = $1392.90
Remaining materials:	$1576.55
Labour:	$2132.50
Total quote for B:	$5101.95

∴ quote from Builder B is cheaper but only by approximately
$50, so it would be which builder was available to do the job
when needed and whose references were better.

Achievement Standard 91030

Area to be painted:

side walls $(3 + 1)$ x 8 + 1 x 2.5	= 18.5 x 4	
	= 74 m^2	
ends (3 x 5 + 5 x 1)	= 20 x 4	
	= 80 m^2	
sloping edge of pool	= 22 + 82	
	= 8.25	
area of bottom	= 8.25 x 5 x 2	
	= 82.5 m^2	

Total area to be painted x 2
(74 + 80 + 82.5)   **= 236.5 m²**

Paint (3 x 10 L @ $185)	= $555
Water blaster	= $90
Concrete (sides 8 x 1 x 2)	= 16 m^2
(end 7 x 1)	= 7 m^2
(circular end (pR2 - pr2)	= 38.465 – 19.625)
	= 18.84 m^2
Total area	= 41.84 m^2
Volume of concrete (41.84 x 0.1)	= 4.184 m^3
Cost of concrete (4.184 x $215)	= $899.56
Labour (60 x $35)	= $2100

Total cost of renovations
$555 + $90 + $899.56 + $2100   = $3644.56

Volume of the pool:	
(bigger part) 2 x 8 x 5	= 80
(smaller part) 2.5 x 1 x 5	= 12.5
Total volume	= 92.5 m^3
Number of litres	= 92 500 L
Number of KL	= 92.5 KL
Cost (92.5 x $13)	= $1202.50

∴ **Total cost of renovations and water = $4847.06**

Achievement Standard 91032

Each step is 250 mm high so ten steps needed to get to the
first floor. The staircase will therefore need a floor space of
2 m on the ground floor. Angle of elevation is 51.30°. Student
drawings can be scale drawings, showing ratios/similar
triangles.

Enrichment

14 a

A1		500 000
A2		250 000
A3	297 x 420	125 000
A4	210 x 297	62 500

b 16

ISBN: 9780170238670